MW01519083

# Mathematics, Philosophy, and the "Real World"

**Judith V. Grabiner, Ph.D.**

THE
GREAT
COURSES

PUBLISHED BY:

THE GREAT COURSES
Corporate Headquarters
4840 Westfields Boulevard, Suite 500
Chantilly, Virginia 20151-2299
Phone: 1-800-832-2412
Fax: 703-378-3819
www.thegreatcourses.com

# Judith V. Grabiner, Ph.D.

Flora Sanborn Pitzer Professor of Mathematics, Pitzer College

Judith V. Grabiner is the Flora Sanborn Pitzer Professor of Mathematics at Pitzer College, one of the Claremont Colleges in California, where she has taught since 1985. Previously she was Professor of History at California State University, Dominguez Hills, where she won the institution's Outstanding Professor Award for 1974–1975.

Professor Grabiner received her B.S. in Mathematics, with General Honors, from the University of Chicago, where everybody reads Plato and Aristotle and Kant and thinks they are relevant. She went on to get her Ph.D. in the History of Science from Harvard University. She has taught or been a visiting scholar at a variety of institutions, including the University of California, Santa Barbara; UCLA; the University of Leeds; the University of Cambridge; the University of Copenhagen; the Australian National University; and the University of Edinburgh.

Professor Grabiner's research has focused on the history of mathematics and on its relationship to philosophy, notably on issues involving changing standards of rigor and proof. She is the author of *The Origins of Cauchy's Rigorous Calculus* (MIT Press, 1981); *The Calculus as Algebra: J.-L. Lagrange, 1736–1813* (Garland, 1990); a forthcoming book, tentatively titled *The Calculus as Algebra and Other Selected Writings*, to be published by the Mathematical Association of America (MAA); and many articles on the history of mathematics. She has won the MAA's Carl B. Allendoerfer Award, for the best article in *Mathematics Magazine*, for three different articles. She has also won the association's Lester R. Ford Award, for the best article in *American Mathematical Monthly*, for three different articles.

Professor Grabiner won the Distinguished Teaching Award of the Southern California Section of the MAA in 2002, and won the association's national teaching award, the Franklin and Deborah Tepper Haimo Award for Distinguished College or University Teaching, in 2003.

Professor Grabiner's career reflects the interdisciplinary nature of her field, the history of mathematics. She has long been interested in teaching mathematics to nonmathematicians, wanting to help them see that mathematics is fun, fascinating, and useful. She tries to do this by linking mathematical topics with the students' own interests, especially philosophy and culture. In turn she has learned a great deal from her students, whose knowledge of how mathematics is used in their own specialties—from Leibniz's philosophy to forensic science, from quilting to baseball—has enhanced her appreciation for the many ways mathematics interacts with the world. Her students have also taught her how to explain things more clearly by their insistence that, now that they are in college, the mathematics has to make sense.

# Table of Contents
# Mathematics, Philosophy, and the "Real World"

# Mathematics, Philosophy, and the "Real World"

## Scope:

How has mathematics changed the way people look at the world? Surprisingly, throughout the Western tradition, views about human nature; religion; philosophy; truth; space and time; and works of philosophy, art, politics, and literature have been shaped by ideas and practices from mathematics. The most influential mathematical shapers, geometry and statistical thinking, are quite different from one another and reflect the contrasting ideas of certainty and unpredictability. Euclidean geometry was used by philosophers to argue that truth could be achieved by human thought and that nonmaterial objects could in some sense be known. Later, the disciplines of probability and statistics provided a way of understanding—with precision—those events in the physical and social worlds that seemed to come about by chance and follow no laws at all.

In these lectures, we will see how all of this happened. We will develop enough of the mathematics—key theorems in Book 1 of Euclid's *Elements*, enough elementary probability and statistics to understand the shape of the bell curve—to appreciate its impact. But our goal will be more like that of the historian: to understand and experience the development of Western thought with a focus on how mathematics has helped determine it.

We will begin in the modern world, with a life-and-death example of probability in modern medicine. We will go on to learn some elementary probability theory and its applications to the study of society. We will see how the constancy of murder and suicide rates over time helped move the focus of study to social, as opposed to individual, causes of crime; how, ignorant of causes, people might first conclude that events were due to chance, but change their views when relevant laws are discovered; and how, if human behavior has social causes, people can still argue for the freedom of the human will. Among the thinkers whose ideas we will discuss are Blaise Pascal, Pierre-Simon Laplace, Adolphe Quetelet, James Clerk Maxwell, and Stephen Jay Gould.

Turning from chance to precision, we will see how the model of certainty exemplified by Euclid's geometry affected views of truth and proof. Plato thought geometry's certainty arose from the eternal, unchanging perfection of its objects and argued for an unchanging reality as the only object of true knowledge. Aristotle, though, saw geometry's certainty in its method: logical deductions from self-evident assumptions. Later philosophers accepted that geometry was certain and true and hoped that by looking at mathematics we could tell what properties all certain knowledge had to possess. We will address Euclid's model by examining his original *Elements of Geometry*, not the watered-down versions taught in modern high schools. And we will see how these ideas crucially influenced philosophers like Descartes, Spinoza, Kant, and Voltaire, not to mention Thomas Jefferson's Declaration of Independence.

Even with all this philosophical backup, the self-evident truths of Euclidean geometry gave way, in the 19th century, to non-Euclidean geometries, which produced an alternative view both of perceptual space and of reality. We will look at this revolutionary development and some responses to its implications, focusing especially on Hermann von Helmholtz's argument about the empirical nature of geometry—a challenge to Kantianism—and on Albert Einstein's use of non-Euclidean geometry in his general theory of relativity.

Next, we will examine some of these questions cross-culturally, first contrasting the approach to geometry in the West with that of classical China, and then suggesting some reasons for the differences.

Finally, we will note that all has not been harmonious in the relationship between mathematics and philosophy. We will look at some prominent critics who found mathematicians and their allies to have claimed far too much for the power of their way of thinking, or who found mathematics and mathematical thought cold, unfeeling, and even totalitarian. These "opponents" will include British economist Thomas Robert Malthus; American poet Walt Whitman; British novelist Charles Dickens; and Russian novelist, playwright, and satirist Yevgeny Zamyatin. In this context, we will also revisit the nuanced position of Blaise Pascal.

Throughout the course, we will be interested in the way philosophers, theologians, political figures, artists, and poets appropriated mathematics for their own agendas. But we will also address the way the nature and practice of mathematics shaped those same agendas. The applications of probability and statistics in modern society may be the example with the most obvious contemporary relevance. But, even more important, the course also illustrates the mathematical backbone of the entire history of Western philosophy. All of Western thought, for good or ill, has been consistently in dialogue with mathematics. Both the dialogue and the examples should appeal to those who know mathematics and are interested in its impact on ideas in general as well as those interested in philosophy who want to know more about the central role mathematics has played throughout its history.

# Lecture Nineteen
# Plato's *Meno*—How Learning Is Possible

**Scope:** Now that we know what Greek mathematical definitions and proofs were like, we return to Plato and his ideas about truth and learning, and we investigate more fully how geometric ideas interact with his philosophy. Plato's *Meno* at first glance seems to deal with the question of whether people can be taught how to be good. As we will see, this leads to a dialogue between Socrates and a man called Meno about how to define "virtue" and about what characteristics the definition of any general concept should have. Meno, after finding out that he cannot define virtue, challenges the whole idea that new knowledge can be gained by someone who is ignorant of the matter. Socrates shows that Meno is wrong, and we see this by becoming participants as Socrates uses an example from geometry and a myth about the immortality of the soul. We will ask whether Plato's account of how learning takes place is either philosophically or psychologically plausible. And we will ask why Plato chose a geometric example to make his crucial point.

# Outline

**I.** Plato's dialogue *Meno* is earlier than both Aristotle's logic and Euclid's *Elements*. The *Meno* shows the climate of opinion in Athenian society about definitions; logic; hypothetical reasoning; geometry; and whether it is possible to gain true, certain knowledge.

**A.** The *Meno* opens with Meno asking Socrates whether virtue can be taught, or if not, how it comes to human beings.

**B.** Socrates says that we cannot answer a question like this unless the key term is defined, so he asks Meno to define "virtue."

**C.** Meno thinks that each kind of person has his or her own type of virtue, and he gives examples.

**D.** Socrates asks Meno to find what all these examples have in common.

    **1.** Socrates's question implies that any collection of things with a common name must have a common underlying idea and thus a definition that identifies what they have in common.

**2.** The *Meno* is famous for being about the problem of universal definition, and we can now see why.

**E.** To show that defining something can be done, Socrates defines "shape": It is the limit of a solid. Note that the successful example is from geometry.

**F.** Meno tries again, but winds up defining virtue by using the idea of virtue itself.

**G.** Meno, frustrated, says that Socrates is like the electric fish, which makes a person numb upon touching it. Before Socrates started asking all those questions, Meno thought he did know what virtue was.

**H.** Socrates says that he also perplexes himself with his questions and does not know what virtue is either, but he is willing to collaborate with Meno to find out.

**II.** At this point the dialogue takes a sharp turn away from the question of defining terms and asks how gaining any kind of new knowledge is possible.

**A.** Meno puts it powerfully: How can you go looking for something when you do not know what it is? And, even if you found it, how could you recognize it?

**B.** Socrates responds to this question in 2 ways.

    **1.** He claims the immortality of the soul and says that the soul once knew everything and so, to learn, needs only to be helped to remember.

    **2.** He calls Meno's slave and, by questioning him, helps him "remember" a truth about—what else?—geometry.

**C.** The geometric problem Socrates gives the slave is this: Given a square whose side is 2 and whose area is therefore 4, how to find a line which is the side of a square with area 8.

**D.** The first step is watching Socrates get Meno's slave to realize that he does not know the answer, because if the slave recognizes his ignorance, he will seek to remedy it.

**E.** We demonstrate the Socratic method by asking the same questions, with the same diagram.

**III.** We ask what the implications are for learning, and for the pursuit of truth, of this successful demonstration.

    **A.** We discuss the psychological plausibility of the idea that learning is recollection and how one might jog the memory.

    **B.** We examine Socrates's spirited defense of the human drive to inquire and seek the truth.

    **C.** We ask why Plato chooses an example that is from geometry to make his point.

        **1.** The example is far from trivial, since it is a special case of the Pythagorean theorem.

        **2.** We note that Plato makes a veiled reference to the fact that $\sqrt{2}$ is not a rational number.

## Essential Reading:

*Plato's "Meno,"* 70A–86C.

## Suggested Reading:

Kraut, *The Cambridge Companion to Plato.*

## Questions to Consider:

**1.** There are special ethics courses for professionals in business schools, medical schools, and law schools, and there is a weekly Ethicist column in the *New York Times Magazine*. In light of all this, is our society committed to the position that virtue can be taught? What other examples can you propose?

**2.** Plato has Socrates say (in Jowett's translation), "We shall be better and braver and less helpless if we think that we ought to inquire than we should have been if we indulged in the idle fancy that there was no knowing and no use in seeking to know what we do not know—that is a theme upon which I am ready to fight, in word and deed, to the utmost of my power." What are the implications of this for our idea of the good life as Plato seems to see it? As you see it?

# Lecture Nineteen—Transcript
## Plato's *Meno*—How Learning Is Possible

Hello again. In this lecture, we go back to the Greeks and back to Plato. This time, we're looking at a dialogue of Plato called the *Meno*. Before, we were looking at Plato's overall philosophy and how Plato used examples from mathematics to help make his general points. This time, we'll be involved with actual mathematics.

We come to the *Meno* armed now with a lot of useful equipment: We've got first-hand experience with what Greek geometry actually looked like, with what the Greeks thought a good definition is, and what the rules are for valid hypothetical reasoning.

One historical reminder before we start. Plato's dialogues are earlier than both Aristotle's logic and Euclid's *Elements*. So we can use Plato's *Meno* to see what the climate of opinion was—before Aristotle and before Euclid—about definitions, logic, hypothetical reasoning, geometry, and whether it's possible to gain true and certain knowledge. Plato's dialogue *Meno* begins with a man called Meno, a visitor to Athens. In the dialogue, Meno asks Socrates a question: "Can virtue be taught? If not, how do people become virtuous?" I don't know why Meno wanted to know, but in our society this is an important question.

Medical schools teach doctors medical ethics; business schools teach business ethics; children in religious school study the ethical teachings of their faith tradition. Does any of this work? Can we teach people to act better? If so, how? That's the question. So, Meno asks the question: "Can virtue be taught?" Socrates answers Meno by saying: "Hey, I can't tell you if virtue can be taught unless we understand exactly what we're talking about. What is virtue? Can you define it for me?"

You know enough about Plato dialogues now, so you know Meno is not going to be able to define virtue. Meno thinks he can define virtue, though, so he gives it a try. He says: "It's an easy question."

Meno says: "The virtue of a man is to be able to carry on the affairs of his city, help his friends, and harm his enemies; the virtue of a woman is to manage her household, and obey her husband." And so on. Socrates says: "That doesn't define virtue, Meno. All you've done is to give us a list. But what's the criterion that gets something onto that list? What do all those things on the list have in common?"

Meno tries again. This time he says: "The man who carries out the political affairs should do it justly, because justice is a virtue." But then he has to concede that there are other virtues—like courage, moderation, and so on. Again, he's just giving a list. Socrates still wants to know what the things on the list all have in common.

Plato here is assuming, and this is consistent with his philosophy of the Forms—although he doesn't mention that here—Plato is assuming that when a lot of things have the same name, there's a reason for that: They really do have something in common—they really are special cases of some general thing.

Socrates spends a lot of time explaining this point to Meno—and, therefore, explaining it to the readers of this dialogue. So it's clear that people in Plato's time needed educating about how to define things. In fact, the *Meno* is sometimes called "the place where Plato poses the problem of universal definition."

What's the problem of universal definition? Given lots of things that have the same name, what's the general thing—the universal—to which they all belong? In this dialogue, Socrates doesn't seem to be able to define virtue either, but he does give an example of a good definition. He defines shape. "Shape," Socrates says, "is the limit of a solid body." The example comes, not surprisingly, from geometry.

Meno thinks he gets it now, so he tries again. He says: "Virtue is to desire good things and provide yourself with them." But Socrates says: "You mean, even if you get the good things unjustly?" Meno says: "No, no, you have to do it in a virtuous way." Now Meno is stuck; he has defined virtue in terms of virtuousness—defined virtue in terms of itself.

So the dialogue so far has given us some sense of what good definitions are—and has shown us the kinds of errors you can make in defining things. But on the question: "What is virtue?" we've made no progress, unless you consider that realizing what we don't know is progress. As we'll see, Plato does think that this is progress.

But Meno is really annoyed. He thought he knew what virtue was. He has made public speeches about virtue to great acclaim, and yet after Socrates asks him questions like "If virtue is getting good things for yourself justly, aren't you using 'virtue' in defining 'virtue'?"—questions that draw out the logical implications of what Meno was saying—and in the face of that, Meno now feels ignorant.

So he turns to making personal remarks. He says: "Socrates, you're like an electric fish that shocks people and makes them numb." But Socrates answers: "If I'm like that—I numb myself, too." "I don't know what virtue is," Socrates says to Meno, "maybe you did know once, but now it seems that you don't." "Still, let us work together," says Socrates to Meno, "and try to find out what virtue is."

Here the dialogue takes a crucial turn. They don't know this particular thing—that's clear. So, can we ever come to know anything? Is Socrates's method of questioning purely destructive? Is Socrates's analytical criticism a road to complete skepticism? Is all that it does is show people that their own views can't possibly be right because they logically lead to contradictions? Those are good questions, and Meno in the dialogue puts it even more sharply. He says: "How can you go looking for something when you don't know what it is? Even if you found it, how could you even recognize this?"

Let me make this more concrete. Suppose I tell you that in the room where you are listening to this lecture, there's a very valuable object—even though it doesn't look valuable. Go find it! Just as Meno says: "How can you find it?" You don't know what it is, so you have no idea where to start looking, and you wouldn't recognize it if you fell over it. Doesn't that describe the human condition?

But Socrates gives 2 arguments that this does not describe the human condition. First, Socrates says: "The soul is immortal. The soul once knew everything, had contemplated reality—but has since forgotten. To learn, then, the soul only needs to be helped to remember. Learning is recollection." That's the first argument. But Socrates's second argument is the reason we're talking about the *Meno* in this course. In the dialogue, Socrates gives a demonstration of how one can seek and then find a new truth. The truth to be found is going to be one in geometry.

So in the next part of the dialogue, Socrates asks Meno to call Meno's slave, a young man—his being a slave is essential to the drama. Since he was born and brought up in Meno's household, Meno can attest that this young man has never been taught geometry in his lifetime. Then, Socrates gives the slave a problem in geometry.

Here's a square; its side is 2; so its area is 4. Of course, we can show this by the multiplication principle: 2 feet on each side, $2 \times 2 = 4$ squares, one foot on a side. The problem is this: We would like to have a square with twice that area—that is, an area of 8. How could we construct such a square? How long is the line that would be the side of what he calls the "double square"—the square with twice the area? Socrates asks Meno's slave: "The original square had a side of 2 feet. How long is the side of the square whose area is double that of the original square?"

Confidently, the slave says: "4 feet." That's wrong, but Socrates doesn't say: "That's wrong." Instead, Socrates draws a picture showing what the square whose side is 4 will look like. Now the slave can see that the area of a square whose side is 4 isn't 8, but 16. Now the slave tries again: 2 is too small; 4 was too big—so he says: "How about 3 feet?" Socrates shows him what the square with side of 3 feet looks like. Look at it. The slave agrees, the area of that one is 9, not 8—so 3 isn't right either.

Here comes the mathematical and philosophical heart of the matter. Socrates says: "Alright, the 3-foot line doesn't work, so what line does work as the side of a square with twice the area of our original square?" "Try to tell us precisely," says Socrates, "or, if you can't give me a numerical value, show us the line." But the slave replies: "I do not know."

Socrates turns to Meno and says: "See? At first he didn't know what line would work, but he thought he knew, and he confidently told us. Now he doesn't know, and he knows he doesn't know. Don't you agree that now he's better off? That getting zapped by the electric fish was a good thing for him?" Meno says "Yes." Socrates says: "His not knowing and realizing that he doesn't know is good because it will make him want to inquire. When he thought he knew, he wouldn't have bothered. He could confidently tell everyone what was, in fact, the wrong answer." Meno isn't smart enough to realize that he's being insulted, but we can see that what Socrates says about the slave is exactly what Meno was doing when he thought that he knew what virtue was.

Anyway, Socrates goes on. He's going to ask questions, and the slave is going to find the answer for himself. Watch. Here's the same original square with area 4 feet. "Now," says Socrates, "let's extend the diagram like this, so that we have another square with the same area, and yet another, and yet another." Here are these 4 equal areas. "How much bigger is the area of my big square than the original square?" asks Socrates.

The slave can count, so he says: "4 times as big."

Socrates says: "But we wanted a square that was only 2 times as big, remember?"

"Yes," the slave says, "I remember."

"Let's draw a line from corner to corner here. Let's shade in this piece. What's the area of this shaded piece?"

The slave says: "It's half the original area." He says it. Socrates doesn't tell him.

Socrates then draws these 3 similar lines and says: "Don't these 4 equal lines contain this shaded area?" The slave says: "Yes, yes, they do."

"How many of these shaded triangles are there in the shaded area?"

The slave says: "4." He says it. Socrates doesn't tell him.

"How many square feet are there in the area of one of these shaded triangles?"

The slave has already said that it's half the square whose area is 4, so he says: "2." He says this. Socrates doesn't tell him.

"So, how many square feet are there in this whole shaded area?"

Now you'd think the slave is getting pretty excited: "8! 8 square feet," he says. Again, he says this. Socrates didn't tell him.

"Okay, this shaded area is a square," says Socrates, "and what line is the side of this square?"

The slave replies by actually pointing to it: "This line." He picks out the line himself.

For people just reading the dialogue without seeing the picture, Socrates spells it out: "You mean the line stretching from corner to corner of the square whose area is 4?" The slave says: "Yes."

Now Socrates introduces some technical terminology. He says: "The learned people call this line 'the diagonal.' [And so, in sort of a legalistic phrase] Would you, Meno's slave, declare that the square with double area is the square on the diagonal?"

He says: "Certainly, Socrates."

That's Plato's theory of education in action. You don't tell people things— you ask questions that enable them to come to the answers on their own.

Socrates says: "This young man got the knowledge out of himself. Nobody taught it to him, but he's got it. So it must have been in his soul already."

We might explain this by telling a story about neural networks in the brain. But what Socrates says is: "The knowledge got into the young man's soul before his soul was put into his body. The soul is immortal—and learning, then, is recollection."

But why in the world would Plato think that this "learning is recollection" thing would be psychologically convincing? Think about how people remind you of things that you might have forgotten. You've probably had a dialogue like this with your family—something like this:

"Hey, do you remember the Jacksons?"

"No."

"Oh, yes, sure you do. They lived in that yellow house down the street."

"No, I don't know"

"They had a black dog."

"I still don't remember."

"Don't you remember the time you were walking by that yellow house and their black dog came out, and barked, and scared you, and you dropped a whole bag of groceries?"

You say: "Oh! Yes! Yes! The Jacksons! Sure! Of course, I remember them." By asking questions based on his knowledge of the thing you've forgotten, the other person has helped you remember. But you had it in your mind all the time. Also, the excitement you feel when you finally do remember the Jacksons is psychologically like the excitement when you see that the square on the diagonal indeed has twice the area of the original square. Ordinary language has a Platonic metaphor for this: "Oh yes, I see the light!"

So, "learning is recollection" is a powerful metaphor, and it gives another justification—besides the metaphor of "turning the soul towards the light"—another justification for Plato's educational theory, and for the Socratic method of questioning.

There's also a moral dimension to what Plato is saying. He has Socrates say this: "I'm not completely sure of every point I've made so far today, but this I'll defend: We will be better and braver if we think we should seek what we don't know, than if we thought that when we don't know we could never find out, so we shouldn't even try." "I will fight for this," says Socrates, "both in word and in deed, with all my power."

Socrates died for that ideal: "Seek to know." The speech I've just quoted from the dialogue the *Meno* is a powerful invocation of the value of seeking the truth. It's a hard ideal to follow.

In the next lecture, we'll see what happens when—in pursuit of this ideal—Socrates and Meno go back to the question of whether virtue can be taught. Now, though, for the purposes of this course, there are a few more things that I need to say.

First of all, to demonstrate his theory of learning, Plato chose an example from geometry. Why? Because his philosophical audience already knew about it? Maybe—I'll return to that point in a minute—but more likely, because in what other subject would we find a piece of pure knowledge that everybody would agree about? Everybody—from an uneducated slave (although he does seem to be able to multiply $2 \times 4$)—from an uneducated slave to Meno, the sophisticated visitor to Athens from Thessaly—everyone, after the right questions are asked, gets this "truth."

When I do Plato's argument in the *Meno* in my class "Mathematics for Non-Mathematicians," I do it Socrates's way. I go through the same argument, drawing the stuff on the blackboard, asking the same questions, and doing it slowly enough until virtually all the members of the class discover the answer for themselves. I can tell it's working, because the students say things like: "Oh! Oh, yes! I see! I see!"

So this method of teaching this particular piece of mathematics really does work. In fact, in most good mathematics classes, it's standard practice to ask questions, to help the students discover the answers for themselves. From the student's point of view, what you are taught this way, you really remember. There's a lot to be said for the Socratic method. As I've said before, Plato's influence is very much with us—not least in the teaching of mathematics.

But back to the specific example from the dialogue—there's lots of mathematics that Plato could have used to do this demonstration. Why this particular example? My guess is that he chose it because of its significance. It's a special case of a really important piece of Greek geometry, the Pythagorean Theorem—that's the theorem that the square on the hypotenuse of a right triangle is equal to the sum of the squares of the other

2 sides. In the special case that Plato uses in this dialogue, the 2 sides of the right triangle are equal—so the square on the hypotenuse is twice the square on one of the sides. The hypotenuse is, as Socrates said, "what the learned call the diagonal." So Plato is saying: "Hey, if learning is recollection, even a slave can learn the Pythagorean Theorem."

One more mathematical point: Remember how, after the slave tried 4 feet and then 3 feet, Socrates asked him: "Try to tell us precisely," and, as I translated it, "if you don't want to give a numerical value, show us the line"—a more precise translation might be "if you don't want to calculate," or "if you don't want to count," "show us the line." Why did Socrates make such a point of this—"if you don't want to give me a numerical value, just show me which line it is"?

Well, what is the precise numerical value of the length of that line? It's got to be a number whose square is 8, and that number involves the square root of 2. $8 = 4 \times 2$. 4 has a square root alright—it's 2. So we could give the precise numerical value of the line if we knew the square root of 2. See, 2 × the square root of 2 × 2 × the square root of 2, regrouping is 2 × 2, which is 4 × the square root of 2 × the square root of 2. The whole point of the square root of 2 is that the square root of 2 × the square root of 2 is 2—so that would give us 4 × 2, which would give us 8 [$(2\sqrt{2})(2\sqrt{2}) = 2 \times 2 \times \sqrt{2} \times \sqrt{2} = 4 \times 2 = 8$].

Okay, so what is the numerical value of the square root of 2? Is it a fraction, like, say, 7/5? No, 7/5 squared is 49/25—that's too small. 50/25 would have been right, but 49/25 isn't. Let's try another fraction. How about 17/12? 17/12 squared = 289/144—that's too big. 288/144 would have been right, but not 289/144—that's too big.

The Greeks didn't have decimal fractions, but suppose they had. I asked my calculator: What's the square root of 2? It told me, to 11 decimal places, the square root of 2 is 1.41421356237. If you square that number, though, you don't get 2. Using a computer with 14 decimal digits, what I got was 1.99999999999999—a bunch of 9's; 14 9's. In fact, there is no fraction—no ratio of whole numbers reduced to lowest terms—$a/b$ such that when you square it, you get 2. The square root of 2 is not a rational number.

Maybe you remember the scene from the Dustin Hoffman film *Rain Man*. Dustin Hoffman plays a man with autism who is a lightning calculator. In that film, a doctor asks him to calculate a square root in his head. So he does it. The doctor then checks the calculation with his calculator, and the calculator agrees exactly with Dustin Hoffman. But that's mathematically wrong. What the calculator has is not the square root. It's just an approximation, accurate to no more decimal places than that particular calculator can produce. All that's happened in the film is that Hoffman and the calculator use the same approximation, same number of decimal places—that's all; neither of them is right.

Unless a whole number is a perfect square—like, say, 4 (which is 2 squared), or 9 (which is 3 squared), or 16 (which is 4 squared)—its square root is never going to be rational, never going to be the ratio of 2 whole numbers. Plato knows this. So why does he have Socrates ask the slave to tell him the precise number—when he knows perfectly well it can't be given? I think Plato's doing this for 2 reasons. First, Plato is showing off a little. Plato's reminding his audience that he knows perfectly well that the Pythagoreans proved that the square root of 2 is not any rational number.

The second—more profound—reason, I think, is that Plato wants to remind us of the method that the Pythagoreans used to prove that the square root of 2 isn't rational. So how did the Pythagoreans (I'm not going to do the details; don't worry) prove that the square root of 2 is not the ratio of 2 whole numbers? My point is this: You can't prove it directly. You can't possibly try every conceivable rational fraction the way I tried to do before—like 7/5, 17/12, 1.414 to 11 decimal places, and so on. Why can't you try every conceivable fraction? Because there are infinitely many such fractions. So you have to do it by indirect proof, by proof by contradiction. You assume there is such a rational fraction reduced to lowest terms, and you show that that assumption leads to a contradiction.

Whether I'm right about Plato's intention or not, this reminder about proof by contradiction helps to prepare us for what's going to come next in the dialogue the *Meno*—namely, Plato's use of hypothetical reasoning and indirect proof. He's going to use those things to try to find out whether virtue can be taught. He's going to take this idea of indirect proof, of proof by contradiction that's exemplified by many things in mathematics—but in particular, the proof that the square root of 2 is irrational. He's going to take that, that piece of intellectually powerful reasoning, and he's going to use that method, exemplified in geometry, in talking about the question that he and Meno were interested in. He's also going to use what Plato in *The Republic* has already told us is the essence of mathematical reasoning, hypothetical reasoning, if-then reasoning—the kind of stuff that we were talking about earlier in the lecture on logic. Plato knows all those things. He's going to take them, and he's going to import them into this dialogue about virtue, and he and Meno are going to use these intellectual tools to attack what is a problem in ethics and a problem in the philosophy of ethics—and the philosophy of politics for that matter. Again, the thing that I keep saying about the intimate relationship between mathematics and philosophy in general and between mathematics and philosophy in Greek culture is really, really, really exemplified in the dialogue, the *Meno*. So we've got all of that ready to come, and that will be the subject of my next lecture on the second part of Plato's dialogue, the *Meno*. Thank you.

# Lecture Twenty
## Plato's *Meno*—Reasoning and Knowledge

**Scope:** Now that Plato has established that learning is possible, he lets Socrates and Meno resume their dialogue about the question "Can virtue be taught?" But since they still have no agreed-on definition of virtue, Socrates addresses the original question by introducing the powerful intellectual tool of hypothetical reasoning. At first he exemplifies this by looking at the implications of an if-then statement about virtue. But in order to further explain how hypothetical reasoning works, Socrates gives a very sophisticated example from geometry. We will explain his reasoning with our own less complicated example, using it to increase our insight into Greek geometry as well as to better appreciate the nature of hypothetical reasoning. We will conclude by following Socrates in explaining the distinction between true knowledge and right opinion, and we will relate it to our earlier discussion of the divided line. Finally, we will go beyond Plato to consider the implications of this distinction for the teaching of mathematics today.

## Outline

I.  Having demonstrated that if we do not know something, we should start researching it, Socrates wants to go back to figuring out what virtue is, but Meno still wants to know whether virtue can be taught even though they do not have a definition of virtue.

   **A.** Instead of concluding that this is hopeless, Socrates suggests using hypothetical, or if-then, reasoning.

   **B.** To explain how hypothetical reasoning works, Socrates gives an example from geometry.

   **C.** We give a simpler example, from Pythagorean number theory, that makes the same point.
   1.  We ask, Is a given number the sum of odd numbers in sequence starting with 1? That is, does the given number equal $1 + 3 + 5 + 7 + \cdots +$ something?
   2.  How, as Meno might ask, can we answer this question if we do not know what the given number is?

3. But it is a fact that the sum of odd numbers in sequence is always a perfect square. For instance, $1 + 3 = 4$, $1 + 3 + 5 = 9$, $1 + 3 + 5 + 7 = 16$, and so on.

4. Using this fact, we can argue hypothetically: If the given number is a perfect square, then the answer is yes, it is the sum of odd numbers in sequence starting with 1; if it is not a perfect square, then it cannot be the sum of odd numbers in sequence starting with 1.

5. So we have answered the question without knowing what the given number is.

D. We explore the power of this hypothetical thinking by examining a classic word problem.

II. Socrates then examines the hypothesis that virtue is knowledge and the hypothesis that virtue is not knowledge.

A. If virtue is knowledge, Socrates says, then virtue can be taught.

B. After much discussion of what goes on in society, Socrates concludes it does not look as though virtue is actually taught.

C. In a brief episode casting light on how Socrates came to be condemned to death by his fellow citizens, in part because of his insistence in questioning the established values of his society, he invites another man into the dialogue: Anytus, who later became one of Socrates's accusers.

1. Socrates asks Anytus to recommend a teacher of virtue for Meno and asks particularly about the Sophists—a group of rhetoricians and intellectuals.

2. Anytus replies that any ordinary citizen of Athens, taken at random, would do a better job teaching Meno virtue than the Sophists would.

3. Socrates asks whether great men of virtue in Athens, like Themistocles, had sons who were also good.

4. Socrates observes that these good men educated their sons in many areas and would certainly have taught them virtue if they could have done so.

5. Anytus, instead of conceding that virtue cannot be taught, tells Socrates to watch out, saying that in Athens, it is easier to do harm to people than to benefit them.

6. Plato is dramatically telling us that Socrates's championing of the life of inquiry and his willingness to fight for the search for knowledge involve real risks.

**D.** Socrates now tells Meno that since there are no teachers of virtue, there are also no students of virtue, and therefore it does not seem that virtue can be taught. So, virtue cannot be knowledge.

**III.** Since there are good men, though, and since it seems that their goodness does not arise from their knowledge, Socrates now asks how people nonetheless can be good.

**A.** Socrates introduces the notion of right opinion: that is, being right about something even though one does not have knowledge of it.

**B.** Although opinions can be wrong, right opinion is correct, by definition.

**C.** Now Meno, who still has not fully grasped what is involved in the search for truth, wonders why knowledge should ever be preferred over right opinion.

**D.** True opinions, says Socrates, need to be fastened in the mind by the tie of cause to become stable like knowledge.

**E.** Thus knowledge differs from right opinion because in the case of knowledge, we know why.

**F.** Virtuous men have right opinion, so they always choose the good, but they do not know why and therefore cannot teach their virtue to others.

**G.** Socrates concludes that virtue is neither natural nor taught but is given to the virtuous person by divine dispensation.

**H.** But, Socrates says at the close of the dialogue, although we have drawn a conclusion, we cannot know the answer for sure until we know what virtue actually is.

**I.** To put it another way than does the *Meno*, hypothetical arguments, though valuable, can take us only to the third level of the divided line, and the knowledge of the form of virtue still lies beyond.

**IV.** We look at what the dialogue tells us about the teaching of mathematics, as considered by the Greeks and as practiced in the present day.

**A.** The idea that a logical proof explains "why" is exemplified by the Pythagorean proof that the sum of the odd numbers starting from 1 must be a perfect square, thus turning the right opinion produced by the assertion of that fact into the mathematical understanding produced by the proof.

**B.** More mundanely, we ask whether electronic calculators know the multiplication they perform or merely have right opinion.

**C.** Finally, we contrast a student who has memorized the formula for 4 choose 2 with someone who knows why the answer must have this form.

**D.** We argue that emphasizing the way mathematics is like philosophy by focusing on "why" enhances both these subjects' drive toward certainty and truth.

### Essential Reading:

*Plato's "Meno,"* 86C–100C.

### Suggested Reading:

Klein, *Commentary on Plato's "Meno."*

### Questions to Consider:

1. The great philosopher Alfred North Whitehead once said, "Civilization advances by extending the number of important operations we can perform without thinking." How can this apparent defense of right opinion be reconciled, if it can, with the strong Greek preference for thinking about "why," so strongly reinforced, historically, by the continuing emphasis on proof in mathematics and the sciences?

2. Should mathematics be taught by the Socratic method of asking questions and letting the students discover (or recollect?) the mathematics by themselves? Defend your answer.

# Lecture Twenty—Transcript
## Plato's *Meno*—Reasoning and Knowledge

Hi. We're going back to Plato's dialogue, the *Meno*. The story so far: Socrates has eloquently persuaded us that if we don't know something, we should try to find out—and we can find out.

Socrates wants to go back to figuring out what virtue is. But that's not what Meno is interested in. Meno still wants to ask whether virtue can be taught, even though they don't have a definition of virtue. Can they, nevertheless, get somewhere with that question: "Can virtue be taught?" "Yes," says Socrates, "I know another way to get a handle on the 'I don't know what it is, so how can I inquire' problem, and that way is let's use hypothetical reasoning—"if-then" reasoning. Socrates, to make this point, gives an example from geometry, but the example is unbelievably complicated. I'm going to explain his point using a different example—really because his is too hard and would lead us astray.

In the process of explaining Socrates's point with my example, I'll also be able to show you a really neat example of Pythagorean number theory—that's another predecessor of Euclid's work. So, it will be cool.

So here's how hypothetical reasoning could solve a problem about something, even though we don't know what the something is. My example is about adding up all the odd numbers, starting with 1—odd numbers 1, 3, 5, 7, and so on. Suppose somebody asks: "Can a given number be written as the sum of odd numbers in sequence starting with 1?" That is, can the given number be written in the form: $1 + 3 + 5 + 7$ and so on, up to the last odd number in some sequence?

"How," as Meno might ask: "How can we possibly answer that question, since we don't know what the given number is?" Actually, we can answer that question. Watch. It is a fact that the sum of odd numbers in sequence is always a perfect square—a perfect square being a number that is a different whole number squared. Let's look at some examples: The sum of odd numbers in sequence is always a perfect square. 1 is 1 squared ($1^2$). $1 + 3 = 4$, and that's 2 squared ($2^2$). $1 + 3 + 5 = 9$, and that's 3 squared ($3^2$). $1 + 3 + 5 + 7 = 16$, and that's 4 squared ($4^2$).

Let's jump ahead a little. $1 + 3 + 5 + 7 + 9 + 11 + 13 + 15 + 17 + 19 = 100$, and that's 10 squared ($10^2$), and so on. So, it looks as though if the given number is a perfect square, then it can be written as the sum of odd numbers in sequence starting with 1. That's a fact. If it isn't a perfect square, then it can't.

Now we can answer the original question: "Can a given number be written as the sum of odd numbers in sequence starting with 1?" We can answer it with this hypothetical statement: If the given number is a perfect square, then it can. You give me a number, say, 16, and I say: "Yes, that's a perfect square—therefore we can do it: $1 + 3 + 5 + 7 = 16$."

You give me another number, say, 14, and I say: "Well, it's not a perfect square, so we can't do it. The sum of the first 3 odd numbers of the sequence gave us 9, adding the next one gave us 16—so for 14, it can't work. So we have completely answered the question: "Is a given number such a sum?" without having to know what the given number is. Hypothetical reasoning has solved the problem. If the given number is a perfect square, then we can do it.

I like my example, but I did something very un-Platonic. I just told you a fact—that the sum of the odd numbers in sequence is a perfect square. I didn't let you discover it for yourself. Worse, I didn't even explain why it's true. Later on in this dialogue, Plato will have something to say about people who do what I did. But I'll rectify my error near the end of this lecture. I will explain why the sum of the odd numbers in a sequence starting with 1 is a perfect square. I'll show you the Pythagorean proof.

Meanwhile, though, let's go back to Socrates's point about the power of hypothetical thinking. Hypothetical thinking in mathematics is not limited to geometry. For a modern instance, let's talk about how we solve problems in algebra by using $x$ for the unknown. Here's one of these classic word problems: These are part of your cultural heritage, too. John is twice as old as Mary. Ten years ago he was 3 times as old as Mary. How old is Mary? How old is John? This isn't a real world problem because if you're going to make this a real-world problem what you would be doing is you'd work out these complicated relationships between the ages, and you'd forget the ages but remember the relationships—this does not happen. But it's a good exercise in algebra.

When you're solving word problems, the first thing you're supposed to do is name what you don't know—so let $x$ equal Mary's age. Here's the point: We don't know what $x$ is. But if we knew what $x$ was, then we could find John's age by multiplying $x$ by 2. Here is the key algebraic step: Assume that $x$ is Mary's age, and if $x$ is Mary's age, then 2 times $x$ is John's age—hypothetical statement, if-then.

Here's another hypothetical statement. We want to know how old John was 10 years ago, because of the conditions of the problem. Hypothetical statement: If John's age now is $2x$, then 10 years ago his age was $2x - 10$. How old was Mary 10 years ago?

Another hypothetical statement: If Mary's age now is $x$, then 10 years ago her age was $x - 10$. We still don't know how old either one of them is, but we're told, remember, that 10 years ago John was 3 times as old as Mary. So, again hypothetically, John's age 10 years ago is going to be 3 times Mary's age 10 years ago. So we have $2x - 10$ (John's age 10 years ago), 3 times Mary's age 10 years ago.

But if this is true, then we can use the rules of algebra to transform this equation into one with an $x$ on one side and with only numbers on the other.

First, let's multiply out the right-hand side 3 times that $x - 10$, and that gives us $2x - 10 = 3x - 30$. Then, we can add 30 to both sides of the equation and subtract $2x$ from both sides of the equation, and that gives us $20 = x$.

Here's the last hypothetical statement: If $x$ is Mary's age, then $x = 20$. That's the right answer. We can check it. If Mary is 20, then John is 40—twice as old as Mary. If their ages are now 20 and 40, 10 years ago she was 10, and he was 30; that's 30—3 × 10. So the problem is solved.

So we see that much of the problem-solving power of algebra comes from hypothetical reasoning. See, we don't know what the number we want is, but if we knew it, then we could multiply it by 2, or subtract 10 from it, and so on, and that's what we do. It is no accident that the person who developed general symbolic algebra in the late 16$^{\text{th}}$ century, François Viète (*Vieta* in Latin) was well versed in Greek geometry. Viète—a really important and interesting guy.

Viète explained what he's doing in hypothetical terms: Assume that you already have the answer (it's not true, but if you already had the answer), you'd give it a name using a letter, and you'd treat it as if it were a known number—so you can double it, or subtract 10 from it, and so on.

So hypothetical reasoning is at the heart of algebra; it's at the heart of modern mathematics as well as ancient mathematics. These were my examples—I mean Plato doesn't obviously know symbolic algebra, but these examples are completely consistent with the point that Plato makes in his more complicated example. He used a mathematical example of a hypothetical argument to get his audience ready for a hypothetical argument in philosophy.

Now we're ready to go back to Plato. "Alright," Socrates says to Meno, "now that we understand about hypothetical reasoning, let's investigate our original question hypothetically. Can virtue be taught? Well, let's make a hypothesis about virtue. Let's make the hypothesis that virtue is a kind of knowledge. After all, knowledge is a good thing; it's good so it's at least a hint we might be looking in the right direction."

We know that knowledge can be taught—"Hey," Socrates might say, "I just taught your slave the Pythagorean Theorem!" So here is Socrates's hypothetical statement: If virtue is knowledge, then virtue can be taught. Okay? So, Meno's question then has an answer.

How are we going to test the hypothesis that virtue is knowledge? Socrates does something that I talked about in Lecture Eighteen, when we talked about if-then logic. I'm going to put his argument, the argument he's going to use in this dialogue, in the form that I used in that lecture. So, let's make that hypothetical statement whose form is "If $p$, then $q$" the beginning of an argument.

"If $p$, then $q$. Not $q$. Therefore, not $p$." "If virtue is knowledge, then virtue can be taught. It turns out virtue can't be taught? Then virtue isn't knowledge." This is how we're going to investigate the question. The argument—"If $p$, then $q$. Not $q$. Therefore, not $p$"—that argument form has a valid form, a truth-preserving form. So, if it turns out as a matter of fact that there are no teachers of virtue and there are no learners of virtue— nobody is teaching virtue to anyone—then it looks as though "Not $q$," "virtue can't be taught," is right. If that turns out to be the case, we'll be forced to conclude that "virtue isn't knowledge." So the thing for them to do now is to see whether virtue is, in fact, being taught or not.

Let's follow Socrates and Meno as they investigate this question: "Is virtue being taught?" It turns out that there's another person around in this dialogue. That person was named Anytus. Anytus was one of the Athenian politicians who helped restore the democracy of Athens after the Peloponnesian War, but I'm afraid that Plato does not think that Anytus was one of the good guys. Anytus was one of the chief accusers of Socrates in the trial that led to Socrates's death. But he's a man with some prestige—so, given that this is a drama, it seems plausible that Socrates would ask Anytus to help him and to help Meno solve their problem.

So Socrates starts talking with Anytus. Socrates says: "Anytus, if we wanted to teach Meno to be a good doctor, wouldn't we send him to doctors?" "Of course," says Anytus. "And to be a good shoemaker, wouldn't we send him to the shoemakers?" "Well, yes." "And aren't there teachers of the arts, like, say, playing the flute, and if Meno wants to learn to play the flute, shouldn't we send Meno to somebody who has set up shop as a teacher of flute playing?" "Of course," Anytus agrees.

Alright. Suppose Meno wants to learn to be virtuous. Socrates asks Anytus: "Shouldn't we send Meno to those people who have set up shop to be teachers of virtue?"

Anytus is mystified. He asks: "Who are these people who are running schools on how to be virtuous?"

Socrates replies: "Well, how about the Sophists?" Since ancient times, the Sophists have gotten rather a bad press. The word "sophistry" has come to mean a deceptive argument.

But at the time of Socrates, the name "Sophists" referred to a group of intellectuals who taught courses in "excellence" or "virtue," who taught rhetorical skills, and who speculated about various philosophical questions. The Sophists certainly helped teach a lot of people how to argue. Anytus didn't like these guys at all. So after Socrates suggests sending Meno to the Sophists to learn virtue, Anytus says: "No way!" What he actually says is: "May such madness not seize any of my family and friends as to go to them!" He thinks they corrupt people.

"Gosh," says Socrates, "have the Sophists done something bad to you?" "No," says Anytus, "I've never had anything to do with them, and that's how I like it."

That last quotation from Anytus shows us where Anytus is on the path from discovering that you don't know and then actually finding the answer. Anytus is still at the beginning: "My mind's made up; don't confuse me with evidence."

But Socrates goes on pursuing his own goal: "Alright, Anytus, so the Sophists won't teach Meno to be virtuous. So tell Meno to whom he should go to learn virtue." Anytus replies: "Any Athenian gentleman he should happen to meet."

Even this doesn't discourage Socrates—because, after all, Anytus is still imagining that someone can teach virtue. So Socrates now asks: "How did these people learn enough virtue to teach it?" Anytus says: "They learned from those who were gentlemen before them, or" Anytus challenges Socrates, "don't you think there have been many good men in this city?"

"Yes, there have been," says Socrates, "but were they able to help others become good?" Socrates now gets down to individual people. For instance, Socrates mentions Themistocles, the Athenian hero of the war against the Persians. "Themistocles educated his son in many areas," says Socrates, "for instance, he wanted his son to be a good horseman, and he sent his son to first-class teachers of horsemanship—and these teachers succeeded. The son was really good. He could ride on horseback standing up and throw javelins while he was doing it. So, of course, Themistocles would have had his son taught to be virtuous if he could have done so. But was Themistocles's son a good and wise man like his father?" Anytus concedes that he was not.

Socrates gives several other examples of good men in Athens—including Pericles, Aristides, and Thucydides—and Anytus agrees that their sons weren't really good men either. "I fear," Socrates says, "that virtue may not be something teachable."

But Anytus changes the rules here. Instead of accepting that Socrates has proven "not $q$" and so they must conclude "not $p$," Anytus makes a threat: "You're too quick to speak badly of people, Socrates. You'd better watch out. It's easier to do harm to people than to benefit them—especially here in Athens." It's Plato who is writing this, and Plato knows that this is no empty threat. There were people in Athens who were out to get Socrates for always questioning their pretensions to know what is right. Anytus was one of them, and they did get Socrates. This threat marks the end of Anytus's participation in this dialogue.

But Plato always portrays Socrates as committed to the life of reason, and this dialogue is no exception. So, threat or no threat, Socrates goes on discussing virtue with Meno. Socrates and Meno eventually agree that the Sophists don't claim to teach virtue, but only how to speak skillfully. They also agree that it doesn't seem as though there's anybody around who claims to be a teacher of virtue. "No teachers," says Socrates, "means no learners. Now a subject with no teachers or learners isn't teachable. So, virtue can't be taught." So virtue can't be taught—virtue isn't knowledge. That's that. We don't know what it is, and we don't know how we get it.

But that is not the end of the story. "You know," says Socrates, "we've neglected something. Knowledge isn't the only thing that directs people to act in a way that reaches their goals."

Meno doesn't understand this, so Socrates gives an example: "Suppose somebody wants to know the way to a nearby city. If you've been there, you know, and so you can give that person correct directions." "Of course," says Meno. "But," Socrates goes on, "what if you've never been there, but you have a right opinion about which road goes there, and you tell the person: 'That's the road to take.' Wouldn't that direct the person rightly also?"

"Yes," Meno agrees.

So Socrates says: "Okay, right opinion is just as good a guide to action as knowledge is."

Meno doesn't get this at first. Meno says: "Look, opinion can be wrong, while knowledge is always right." But Socrates says: "I said 'right opinion.' That's always right. If somehow somebody's opinions were always right, he'd be just as good a guide as the person who knew."

Knowledge, of course, is a lot of work—so at this point Meno really likes the idea that there might be a substitute. He says: "Right opinion seems pretty good. I don't even see why people should prefer knowledge." But, of course, Socrates isn't going to accept that. So Socrates gives his view about the difference between knowledge and right opinion. "Right opinion isn't stable," he says, "but knowledge is bound by the tie of cause."

What Socrates means here is that with right opinion, the person knows that something is true. But with knowledge, the person not only knows that it's true, but knows why it's true—and if you know why it's true, then you can teach the thing to others. If all that you have is right opinion, you can always choose what's right, but you can't explain what you know. If all you have is right opinion, you can't do what Socrates did with the slave—you can't ask the right questions that help others discover the truth for themselves.

"These people like Themistocles, they're good, and they're just, and they can run a city well, but they don't have any more understanding of what they're doing," says Socrates, "than do soothsayers or inspired diviners—so they can't teach virtue, not even to their own children." So, for the original question of whether virtue can be taught, or, if not, where virtue comes from, the answer at this point in the dialogue is that it isn't something that can be taught: Virtue comes by divine dispensation to those who have it. It's not knowledge. It's right opinion. That's really the end of the dialogue, except that Socrates points out they can't be sure they've got the answer, because they still don't know what virtue is. Or, to put it another way than the way Plato puts it here—hypothetical arguments, though valuable, take us only to the third level of the divided line, and the knowledge of the Form of Virtue still lies beyond.

I want to go beyond Plato's dialogue, though, and look at the difference between right opinion and knowledge—focusing on mathematics. Aristotle (you'll recall after Plato) thought that a proof ought to explain cause and effect—it ought to explain not only that the thing was, but why it was. Plato didn't say this in the same way, but he seems to agree. So let's go back to the example of the sum of the odd numbers starting with 1 being a square number and see whether so far we have knowledge of it— or just right opinion.

I showed you a few examples, and then I said that it was always true that the sum of the odd numbers starting with 1 gives you a square. Probably you trust me not to mislead you, so you believe this. But at this stage, unless you've seen a proof before, it's still just right opinion. You'll always get it right, but why is it true? Let's look at the actual proof, the why, as given by the Pythagoreans. The Pythagoreans liked to represent numbers with pebbles or dots. What's an even number? It's divisible by 2—so it can always be represented by 2 equal lines of pebbles, like this: 2 identical columns of pebbles; that's the even number 8.

Or, like this: 1 column and 1 row, but each with the same number of pebbles. That's how you'd represent an even number. That's also 8. An odd number is 1 more than an even number. So we can represent it as we just represented the even number—by adding 1 more pebble at the top-right in a sort of upside-down, L-shaped arrangement. That's the odd number 9: It's 2 × 4 + 1. Got that? Let's draw a picture like this of the first odd number, 1, and then add the next odd number to it, and then add the next odd number to that, and so on—geometrically. The first odd number, pictured here, is 1. That's not very interesting. The next odd number is 3. That's 2 pieces—a row and a column of 1 pebble each, and then the extra 1 in the corner.

Hey, that looks very much like a square. That's why they're called square numbers. You get a picture—it's a square: 1 (the original 1) + 3 (this upside-down, L-shaped thing) = 4; it's a square whose side is 2. Let's add the next odd number, which is 5—that's 2 here, 2 there, plus 1 (2 × 2 + 1).

Again, we add a little row of 2 on the top, a little column of 2 on the side, and the extra 1 in the corner, and it looks like this picture, and so on. See, each time we add another odd number, what we do is we add another L-shaped piece—2 times the side of the previous square plus 1. So each time we add another odd number, we get another square whose side is 1 bigger—that's the next square.

A brief aside for those of you who prefer algebra to geometry. Everybody else just kind of tune out for about 10 seconds. The first odd number, 1, is a square. We have in general that $n$ plus 1 squared is $n$ squared plus $2n$ plus 1 $[(n + 1)^2 = n^2 + (2n + 1)]$—you expand out the left-hand side, you get the right-hand side. That shows that you can always get the next square by taking the previous square, $n$ squared $(n^2)$, and adding to it the odd number composed of 2 times $n$ plus 1 $(2n + 1)$.

Back to geometry. Everybody, you can come back. This visual proof is just lovely, and it really does explain why the sum of the successive odd numbers always gives you squares. Once you've mastered this proof, you will have knowledge—not just right opinion—about this topic. I've been teaching mathematics for many years, and I am absolutely sure that Plato is right about this. You can't teach mathematics if you don't know the "why." I ask my students: "Can a calculator teach you how to do multiplication?" The answer is: "No."

See, any time you give it 2 numbers to multiply, the calculator always gives you the right answer. But the calculator only has right opinion. It's always right, but it doesn't know why—it doesn't know anything; it's just following instructions of its programmer; it's like divine dispensation. So it can't teach you anything.

Last year I gave a talk to some elementary school teachers about why it's important to teach the kids why things work—rather than just how to work the algorithms. In that talk, I quoted a little satirical poem—it's the one I quoted in our introductory lecture—about formulaic teaching that goes: "Ours is not to reason why; just invert and multiply." I said, look, that lets the kids solve individual problems all right, but when they get to algebra, they can't cope with algebraic fractions—complicated ones—if they don't understand why the rule for dividing fractions works. Just memorizing a rule or a formula is not enough.

Here's another example from this course. Remember how we figured out what "4 choose 2" is from the multiplication principle? The world is full of people who if you ask: "Hey, what's '4 choose 2'?" would say: "Oh, I memorized a formula for that! I can do it!" But just memorizing the formula and applying it to this particular example gives you only right opinion.

If somebody says: "Why does that formula work?" you couldn't explain why it worked to anybody else. But we know why, right? "4 choose 2": From the multiplication principle, there are 4 ways to choose the first of the 2 objects, and 3 ways to choose the second—so that's $4 \times 3$ total ways of picking those 2. But wait, we've counted every separate ordering of the members of each pair, and we want to count each pair only once—so we divide by the number of ways to put 2 objects in order, which is 2; so the answer is $4 \times 3$ divided by 2, which is 6 ($4 \times 3 \div 2 = 6$).

That's why the answer is what it is—and since we know why, we can use this knowledge to teach other people stuff or to solve a range of other problems, as we did back in Lecture Eleven. Plato cared a lot about mathematics and teaching mathematics, as we saw in discussing *The Republic*. In my opinion, Plato tells us more about the best ways to teach mathematics than any other philosopher.

In the next lecture, we'll return to Euclid and see a different set of examples of the use of logical argument in the service of truth. For now, though, let me make one last point: Mathematics, in many ways, is like philosophy. This is not just because they both use logic. It's even more important, I think, that people in both fields—mathematics and philosophy—care deeply about answering the question: "Why?" Proofs, whether in philosophy or in mathematics, answer the question: "Why?"—and take us from right opinion to knowledge. Thank you.

# Lecture Twenty-One
# More Euclidean Proofs, Direct and Indirect

**Scope:** We now return to Euclid's geometry, with the long-term goal of knowing the key theorems he needed to establish his logically elegant and philosophically important theory of parallels. We will treat that theory in Lectures Twenty-Five and Twenty-Six. Now, we will look at some of Euclid's theorems about the angles formed when 2 straight lines intersect at a point. Euclid takes great care to show the reasoning behind every step of Proposition 13, though some of its assertions appear to be obvious. He even ensures that his wording is ideal and not dependent on the physical world. We then look at the converse—Proposition 14, about intersecting straight lines—which requires an indirect proof. Proposition 15 is the first example in the *Elements* of a theorem that is not completely obvious from the diagram. We will go through Euclid's beautifully clear proof of it. Finally, we will go through a complicated direct proof about triangles—Proposition 16—that plays a crucial role in the theory of parallels. Throughout this lecture, we will focus on the logical structure of this part of Euclid's *Elements* by emphasizing which of the basic assumptions about geometry—the assumptions Euclid calls postulates—are used in each proof.

# Outline

I.  We now move from philosophy back to geometry, where we will observe the same important role given to definition and to logical argument leading to truth and certainty, but with a different subject matter.

    **A.** The particular piece of geometry we will focus on is Euclid's theory of parallel lines.

    **B.** The propositions in Euclid's *Elements* most needed for this theory involve the angles formed when 2 lines meet at a point.

II. We look at Propositions 13–16 of Book 1 of the *Elements*.

    **A.** To proceed with Proposition 13, we will need Definition 10, which says that when a straight line is set up upon another, making the adjacent angles equal, the equal angles are called right angles.

**B.** Proposition 13 says that if a straight line set up upon another line makes angles, it will make either 2 right angles or angles whose sum is 2 right angles.

   **1.** The proof is essentially a cut-up-the-total-angles exercise.

   **2.** It rests, crucially, on the axiom about equals added to equals.

   **3.** Euclid takes care to refer to the sum of the angles as equal to 2 right angles, instead of 180°. This preserves the ideal nature of the proof.

**C.** Proposition 14 is more interesting; it is the converse of Proposition 13.

   **1.** A proposition with the form "If $p$, then $q$" has as its converse the proposition with the form "If $q$, then $p$."

   **2.** If a proposition is true, its converse can be true or false; if it is true, it requires proof.

   **3.** Proposition 14, the converse of Proposition 13, says that if we have 2 adjacent angles equal to 2 right angles, then the straight lines making up the outside of those angles lie in the same straight line.

   **4.** We give Euclid's indirect proof of Proposition 14; the proof requires the truth of Proposition 13.

**D.** We turn to Proposition 15, arguably the first proposition in Euclid that is not immediately obvious from the diagram.

   **1.** It states that when 2 straight lines intersect, the vertically opposite angles formed are equal.

   **2.** We prove Proposition 15; it is a direct proof, relying on the axioms, on Postulate 4, and on Proposition 13.

**E.** Proposition 16 states that if the base of a triangle is extended so that the base and the adjacent side form an angle (called an exterior angle), the exterior angle is greater than either opposite interior angle.

   **1.** The proof requires Propositions 3, 4, and 10. We will give a brief overview of these propositions, though we will not prove them.

   **2.** Later on when we talk about non-Euclidean geometries, it will become important to notice the essential role Postulate 2 plays in this proof, since it is Postulate 2 that allows us to extend the base of the triangle to any desired length.

**F.** Now that we have this collection of proved propositions, we have, as we will see in Lectures Twenty-Five and Twenty-Six, established the basis for Euclid's theory of parallels.

**III.** What we have just done reminds us once again of the basic facts about demonstrative sciences in general, as well as Euclid's *Elements* in particular.

    **A.** Assuming the truth of the basic principles, we proceed by logical deduction to prove a wealth of results.

    **B.** As Isaac Newton, a 17<sup>th</sup>-century master of demonstrative science, so beautifully said, "It is the glory of geometry that from so few principles it can accomplish so much."

### Essential Reading:

Heath, *The Thirteen Books of Euclid's Elements*, vol. 1, Book 1, Theorems 13–16.

### Suggested Reading:

Katz, *A History of Mathematics*, chap. 3 (chap. 2 in the brief edition).

Knorr, *The Evolution of the Euclidean Elements*.

### Questions to Consider:

**1.** If there were some surface on which Postulate 2 were false, would this mean that Proposition 16 was necessarily false, or possibly just that Euclid's proof was insufficient?

**2.** If there were some surface on which Proposition 16 were found to be false, but where all the axioms, propositions, and postulates except for Postulate 2 were known to be true, would Postulate 2 necessarily have to be false on that surface?

### Answers:

**1.** "If Postulate 2, then Proposition 16" does not necessitate "If not–Postulate 2, then not–Proposition 16." This would be the fallacious argument "If $p$, then $q$. Not $p$; therefore, not $q$."

**2.** "If Postulate 2 and other things, then Proposition 16" does necessitate "If not–Proposition 16 (and if all the "other things" are still true), then not–Postulate 2." This would be the valid argument "If $p$, then $q$. Not $q$; therefore, not $p$."

# Lecture Twenty-One—Transcript
# More Euclidean Proofs, Direct and Indirect

Welcome back. We now move from philosophy back to geometry. In geometry, as in philosophy, the leading roles go to definition and to logical argument—but the subject matter, as Aristotle would say, is different. The example of geometry is much more powerful than that of philosophy because, instead of conclusions that people are going to dispute, in Euclid's geometry the conclusions we prove were historically viewed as true—and agreed on for 2000 years by absolutely everybody.

The point of the present lecture on one small part of Euclid's *Elements* is not to teach you geometry—although I'm happy to do that—but to let you appreciate the power of this example of a logical structure of proved truths by actually getting us all inside it. The piece of geometry that has had the greatest philosophical impact is Euclid's theory of parallels—parallel lines. As we'll see in the last part of the course, Euclid's theory of parallels has had immense influence in art, in science, and in philosophy.

The overthrow of Euclid's theory of parallels, with the invention of non-Euclidean geometry, had equally immense influence. So, we'll try to keep our eyes on the prize there—ideas about truth and about space and their impact on thought in general. But the foundation of these ideas about space and truth is in the details in Euclid's geometry. In this lecture, I will begin to lay that foundation.

The propositions that Euclid needs most before he's going to be ready even to start the theory of parallels—those propositions are about the angles formed when 2 lines meet at a point. Parallel lines are lines that never meet. These are about when 2 lines do meet at a point, as in these pictures.

What I'm going to do today is to look at 4 propositions in Euclid that hang together as a little logical unit—Propositions 13, 14, 15, and 16 of Book 1 of Euclid's *Elements*. These propositions—or theorems, again to use the Greek name—these theorems are the ones about the angles formed when 2 lines meet at a point. I'll prove these theorems the way Euclid proved them. We'll start in Euclid's Book 1 with his Proposition 13—and for the proof of that theorem, we're going to need Euclid's definition of right angles. So, let's review that.

When a straight line intersects another straight line so that the adjacent angles formed are equal, those equal angles are called "right angles." So in this picture, if angle A is equal to angle B, they're right angles by this definition. We're also going to need another theorem Euclid proved—I might as well give you its number; it's Proposition 11—that says we can construct a perpendicular—that is, a line that is at right angles to a given line. So accept it—we can construct a perpendicular.

Now, we're ready for Euclid's Proposition 13. It says: If a straight line intersects another straight line, either it makes 2 right angles, or it makes 2 angles whose sum is 2 right angles. Here's the picture. That is, we take the line AB, and we set it up so it intersects the line CD. (Notice, by the way, that the alphabetical order in which we put letters on the diagram again helps us see in what order the lines come into our consideration. That's part of the tradition of Greek geometrical diagrams, and it kind of helps us understand how the proof is going to go.)

The proposition says that either angle ABC and angle DBA—either they're both right angles, or that when you add the 2 angles together, ABC and DBA, their sum is equal to the sum of 2 right angles. Of course, if each of these 2 angles is a right angle, our proof is done; of course, it's 2 right angles. So we're going to consider the situation where the 2 separate angles are not right angles. Then, the goal is going to be to prove that this angle here and this angle here, ABC plus DBA, equals 2 right angles.

Of course, it looks obvious, but that's not good enough for Euclid. He's going to proceed by a series of presumably obvious steps. He wants to prove—give a reason for—each apparently obvious step. Let's watch how he does it. This is essentially Euclid's proof.

Now again, as I said, if angle ABC (angle 3 in my diagram) is a right angle, we're done. Let's assume that angle ABC is not a right angle. In that case, we can construct (look over at the picture) the line EB that does make a right angle with the line DC. (As I said, he's proved earlier, in Proposition 11, that we can do this.) Then angle DBE (angle 1 in my picture) is equal to angle EBC, because that's the definition of right angles.

Now the proof is going to work by cutting up and reassembling the angles that we're interested in. First, focus on this right angle EBC. That's angle 2 plus angle 3. Let's write that as an equation: EBC = angle 2 + angle 3. It's an equation, so we can add the same thing to both sides and the equation will still be true—that's one of Euclid's axioms. So we add the right angle EBD, that's angle 1, to both sides of this equation. On the left-hand side I'm going to call it EBD, and on the right-hand side I'm going to call it angle 1. But we're adding the same thing to both sides of the equation—if equals are added to equals, the sums are equal: Euclid's Axiom 2. So now we get: angle EBC + angle EBD = angle 1 + angle 2 + angle 3.

Now, we're going to focus on a different angle, angle DBA. That's one of the ones that we're really interested in, one of our original angles. That's angle 1 + angle 2. So we write angle DBA = angle 1 + angle 2. Now, we're going to do the same thing, add the same thing to both sides—add angle ABC, which is also angle 3, to both sides of the equation. We'll add in the form ABC to the left side and angle 3 to the right side. We're still adding equals to equals—the sums are equal, so that's legal, Axiom 2. That gives us the angle DBA + angle ABC = angle 1 + angle 2 + angle 3.

Look at these 2 equations that I've anachronistically labeled with Roman numbers I and II. We've got 2 things both equal to the same thing, angle 1 + angle 2 + angle 3. Two things equal to the same thing are equal to each other—that's Euclid's Axiom 1. So, angle EBC + angle EBD must be equal to angle DBA + angle ABC. But angle EBC and EBD—those are 2 right angles, precisely 2 right angles. So if EBC and EBD are 2 right angles, then DBA + ABC also add up to 2 right angles.

That's what we set out to prove, right? It must be true because we proved it. Notice careful and detailed mustering of the reasons for every little thing and good logic; the logic is very important—it's got to be perfect. You may wonder: "Why doesn't Euclid say that the sum of the angles is 180 degrees, instead of saying that it's 2 right angles?" Euclid doesn't say why he isn't using degrees—he never tells us why he's making the choices he's making. But I think it's clear nonetheless. Two right angles—that's completely part of mathematics. But the size of a degree, so that the sum of the 2 angles is 180 degrees, the degree is arbitrary. You might as well talk about lengths measured in inches, or centimeters, or the foot of the king.

Where does measuring angles in degrees even come from? In fact, degrees come from ancient Babylonian astronomy. I mean the Greeks used this Babylonian measure: The Babylonians divided the circle into 360 degrees, the whole circle, 360 degrees—half the circle, 180 degrees. Why did they divide the circle into 360 degrees? Probably because it takes the sun about 360 days to appear to make 1 complete circuit of the earth. It moves through the stars roughly 1 degree a day. Yes, I know the year is 365 and a quarter days, and the Babylonians knew this, too. But they had a base 60 number system—that, by the way, is why we divide the degree into 60 minutes, and each minute into 60 seconds—and for time, too. That's also Babylonian. So the 360-degree circle seems to go nicely with base 60 numbers. Anyway, that's what most scholars think—but it does come from the Babylonians.

But you see the theorem I just proved isn't about anything in the visible world; it's geometry. Arbitrary units—whether they're feet, inches, meters, or degrees—they don't belong here. If you asked Plato: "Hey, how come 2 right angles and not 180 degrees?" I think he might say: "Inches and degrees—those are physical measurements; they belong on the second level of the divided line. It's not where we are. We're up on the third."

Euclid, of course, doesn't say, but I think he'd agree: 2 right angles—that works anywhere in the universe. It's entirely independent of culture and history. All you need is the idea of 2 intersecting lines and the idea of equal angles.

Back to Euclid and to his next theorem: Remember that the one I just proved said that: If 2 straight lines intersect, then the angles add up to 2 right angles. The next theorem is what is called the "converse" of that one.

If we have 2 angles, like CBA and ABD that add up to 2 right angles, then the points C, B, and D must lie on a straight line. Let me make the converse part of it clear. That is, If the angles add up to 2 right angles, what is intersecting here must be 2 straight lines. That's the converse of "if what's intersecting here is 2 straight lines, then the angles add up to 2 right angles."

Remember our earlier discussion in logic about the relationship between "If p, then q" and "If q, then p"? These are called converses of one another. We know that logic by itself doesn't give us the converse. If the converse is going to turn out to be true, it needs its own proof—and Euclid proves it.

I'm not going to do the actual proof in this lecture—we haven't got time to prove everything, and Euclid isn't going to need the converse for the part of the theory of parallels that we're talking about. But I do want to give a real-world application of the converse of Euclid's Theorem XIII, the one I did prove.

I build model ships. Often I cut a tiny little piece of wood to be a rectangular part of the ship's fittings. How can I tell if each corner is precisely a right angle? Because these pieces of wood are so small it's very hard to tell by just looking—it's very hard to tell even if you have another right angle to compare it with. But, suppose you take that angle you've cut together with something of the same size that you already know is a right angle. So, I'm going to do that. I'm wondering whether this is a right angle.

I know that this is a right angle. So I'm going to put them together. If they are 2 right angles—according to this theorem, Theorem XIV—then we've got to have a straight line. Let's have a look. Look along the bottom there. Is that a straight line? Totally not. You can easily see a very, very slight deviation from straightness. That's something we can do very precisely as human beings. So you can tell that this is not a right angle. Conversely (I'm going to do the converse now), if (this is Proposition 13, the one I proved) conversely, if the bottoms do line up in a straight line, I can conclude that I do, indeed, have 2 right angles—which I do. Right?

Now we're ready to go on to Euclid's next theorem—that is, if 2 straight lines intersect, the vertically opposite angles formed are equal. That is, in this picture, angle 2 = angle 4, and angle 1 = angle 3. This theorem is often stated as "vertical angles are equal," but my students reasonably enough say: "Well, if angle 2 and angle 4 are vertical angles, then we should call angle 1 and angle 3 "horizontal angles." So, vertically opposite angles, angles opposite the vertex, the point where the lines intersect—vertically opposite angles are equal.

This may be the first proposition in Euclid that you might not think is immediately obvious just by looking at the diagram. In fact, people who like to teach geometry intuitively, instead of by using proofs, sometimes try to convince their students that this result is true by opening and closing 2 hinged rulers. You see the vertically opposite angle—this one and this one? They stay equal. You get one bigger, the other one gets bigger the same way. Look at the ones on the side, likewise. They get smaller; they get bigger. This theorem, by the way, helps you see why, when you're using scissors, the blades are cutting closer and closer together precisely when you bring your thumb and your fingers very close together. Vertical angles are equal. I'm not going to do that again.

But, of course, moving scissors—that's not a good enough proof for Greek geometry. So let me show you how Euclid's proof goes. It depends completely on the theorem that I proved before—that when 2 straight lines come together, the angles add up to 2 right angles. Do you recall that one? This is a very nice, clear proof. First, let's concentrate just on angles 1, 2, and 3: angle 1 + angle 2 = 2 right angles. Why is that? Because of what we just proved. When 2 straight lines come together, the angles add up to 2 right angles. That's angle 1 and angle 2. Likewise, angle 3 and angle 2 add up to 2 right angles—again, because 2 straight lines have come together, and, therefore, the angles have to add up to 2 right angles.

These 2 sums of angles (angle 1 and angle 2, and angle 3 and angle 2) are both equal to the same thing—2 right angles—and therefore, angle 3 + angle 2 and angle 1 + angle 2 have to be equal to each other. Two things equal to the same thing are equal to each other. So now we have this equation here, and if we subtract equals from equals, the remainders are equal. That's one of Euclid's axioms. So let's subtract angle 2 from both sides of this equation—or, geometrically, here we have angle 1 + angle 2; here we have angle 3 + angle 2—we've got these 2 overlapping 180-degree angles. So, let's subtract this angle in the middle from both of them—from this and from this—and we're left, whether you do it in terms of the equation or geometrically, with angle 1 = angle 3. That's half the theorem. The other half is to prove that angle 2 = angle 4, and you can easily do that in exactly the same way. All right, vertical angles are equal.

Now to the last proposition in this little section of Euclid. This one is by far the most complicated—both in its statement, and even more in its proof. But it's going to be very important—both in Euclid's theory of parallels and, later on, in non-Euclidean geometry. We're going to need a few more of Euclid's earlier theorems that I don't have time to prove for you, but that I will state. So, either trust me, or remember your high school geometry, or look them up in Euclid's *Elements*—I'll give you the references, the proposition number. Anyway, here are the ones that I'm going to use.

Proposition 3: Given a straight line, we can lay off a smaller straight line of a given length on it. Given a straight line—we got another one—we can lay off a smaller straight line of a given length on it. That's just a kind of construction. It seems obvious, but, of course, Euclid proved it—Book 1, Proposition 3. The next one you may remember from high school: They called it "side-angle-side." It's Proposition 4: If 2 sides and the included angle of one triangle are equal to 2 sides and the included angle of another triangle, the triangles are said to be congruent—that is, equal in every respect, angles and lines—side-angle-side equals side-angle-side. Last, Proposition 10 of Euclid: We can bisect a line. That is, we've got a line—we can find the point in the middle that divides that line in 2.

Now, we're ready for the statement and proof of the theorem that I want to prove: Euclid's Theorem XVI. Here we go. In any triangle, ABC, if we extend the base BC to some exterior point D, the exterior angle formed, that's the angle outside the triangle here, angle DCA is greater than either of the opposite interior angles—that's angle A or angle B. Notice either opposite interior angle. Angle DCA doesn't have to be greater than the next-door angle, namely BCA—and, in fact, in my diagram it's not.

You may remember from high school geometry that angle ACD is actually equal to the sum of angle B and angle A. That's a much stronger result. But that result can't be proved with the machinery that we have so far. That result also is going to turn out to need Euclid's Postulate 5. Nevertheless, this result here that I'm about to do is very important. Again, just looking at this picture isn't enough to convince you that it is always true. We're going to need a proof to settle this one. It isn't obvious.

I'm only going to do half of Euclid's proof—that angle ACD is greater than angle A. But the proof that angle ACD is greater than angle B works exactly the same way. Before I do any actual proving here, I'm going to need to make lots of constructions in this original angle ABC. So, I'm going to do it step by step.

Here again is triangle ABC. First, we extend the base to D. Then, we bisect the side of the original triangle—that's the side AC, at point E. Remember, Euclid has already proved that we can bisect a line, so we can do that.

Then, we connect the points B and E—we can always construct a line between 2 given points; that's Postulate 1. Then, we extend the line BE to F, so that BF is sufficiently long so that EF = BE. Who says we can lay off the given length BE on the extension of that line? That's Proposition 3, one of the ones I stated earlier—that given a line and another line, you can lay off the segment here.

But wait a minute. Who says that we can extend the line BE as far as we want—in particular, as far as F? Who says I can extend the line BE to double its length? Postulate 2 says that. Postulate 2 says that a straight line can be extended to any length we want. So this proof that I'm about to do depends on the truth of Postulate 2, that we can extend the line as long as we want. We'll need to recall this fact when we get to non-Euclidean geometry. But let's go on.

Finally in this construction, we connect points F and C with the straight line FC, because we can draw a straight line between any 2 points. That's the completed diagram. Notice again, by the way, how the alphabetical order, labeling the points, helps us keep track of the order we followed in making these constructions. That's a nice fact about Greek geometry.

Now for the proof. The strategy of the proof, the overall strategy, is going to be to break up the larger triangles here into smaller triangles. We're going to prove that the smaller triangles are congruent, equal in all respects, using side-angle-side. Once we have congruent triangles, we're going to have some equal angles for us to reason about.

The way we're going to get the final inequality—that the exterior angle is greater than either opposite interior angle—will be to use Euclid's axiom that the whole is greater than the part. So that's the strategy. Here we go. First, I'm going to prove that triangle AEB is congruent to triangle CEF. As I said, I'll do that using the side-angle-side proposition.

First, the equal sides: AE = EC because we bisected AC to get E—so, of course, they're equal. BE = EF because we constructed the line EF precisely to be equal to BE. So we've got the 2 equal sides. Now for the included angles. The angles AEB and FEC are equal. Why is that? Because they're vertically opposite angles, and I proved earlier (Euclid proved earlier) that vertically opposite angles are equal.

So now we have side-angle-side is equal to side-angle-side—so, by Euclid's Theorem IV, those triangles are congruent. That means they're equal in all respects. So we can conclude that the angles in corresponding positions are equal—that is, in particular angle BAE, or angle A if you like, is equal to angle ECF. But we know that angle ECD is greater than angle ECF, because the whole is greater than the part—that is one of Euclid's axioms. So, since angle A is equal to angle ECF, angle ACD is also greater than angle A—angle BAE if you like—and that's what I told you I'd prove. The exterior angle of a triangle is greater than the opposite interior angle BAE. Euclid proves that angle ECD is greater than angle B or angle ABD in the same way.

Let's sum up what we've learned, geometrically, from each of these 4 propositions. First, if 2 straight lines come together, the angles formed add up to 2 right angles. Second, the converse: If the angles add up to 2 right angles, the lines involved are straight—form a straight line. Third, if 2 lines come together, the vertically opposite angles formed are equal. The proof of this one requires the one that if 2 lines come together, they add up to 2 right angles. Finally, the exterior angle of a triangle is greater than either opposite interior angle. The proof of that needed both of the theorems I proved before and Postulate 2.

When we get to Euclid's theory of parallels in Lectures Twenty-Five and Twenty-Six, you'll see that the collection of propositions I've just proved are the essential basis for that theory. So you'll see it really work out neat. But already, we've seen enough of the logical structure of Euclid's geometry to be able to imagine its impact on philosophers—on people who want to find truths, and who want a method of proving truths.

So, before we jump into Euclid's theory of parallels, I'm going to turn back to philosophy. In the next few lectures, we will look at some major philosophers whose work respects, exemplifies, and extends Aristotle's ideal of demonstrative science, as embodied in Euclid's *Elements*. But before I end this lecture, let me sum up what the proofs we've just gone through illustrate about demonstrative science in general, and Euclid's *Elements* in particular.

Three main points: First, as Plato and Aristotle both pointed out, geometry proceeds by logically valid arguments, but geometry is hypothetical—the truth of the conclusions depends on the truth of the axioms and postulates. You saw that; as I was doing Euclid's proofs, I needed to use various axioms and postulates repeatedly.

Second, even from just this little bunch of 4 theorems, you can see how Euclid's *Elements* is a long chain of logical arguments. Each theorem may be short in itself, but they build on each other to reach more and more complicated things, like "the exterior angle is greater than either opposite interior angle." I think we also can see how beautifully apparently unrelated theorems fit together to produce a finished proof. I mean, who would have expected that we can bisect a line that's going to give us "the exterior angle is greater than either opposite interior angle"?

Third, the method is amazingly powerful. A handful of postulates and axioms produces a wealth of theorems. No wonder Euclid's *Elements* caught so many people's imagination. So I'm going to let Sir Isaac Newton—surely a master of demonstrative science—write the conclusion for this lecture. Newton said: "It is the glory of geometry that from so few principles it is able to accomplish so much." Thank you.

# Lecture Twenty-Two
## Descartes—Method and Mathematics

**Scope:** We now turn to the first, and arguably most important, of modern thinkers decisively influenced by Euclid's geometry. René Descartes is often considered the founder of modern philosophy. We will look at some of his philosophical work, partly to appreciate its seminal nature, but also to see how much ancient geometry shaped his work. As an independent coinventor (with Pierre de Fermat) of analytic geometry, which links algebra and geometry, Descartes had to be a master of both Greek geometry and Renaissance algebra. We will of course look at his famous "I think, therefore I am" argument and his views about God, but we will concentrate on what his famous method for scientific discovery owed to mathematics. We will also examine the extent to which the Euclidean ideal guided Descartes' physics, and in particular how these ideas helped him anticipate Newton's first law of motion.

## Outline

**I.** René Descartes is often called the founder of modern philosophy.

    **A.** He began by repudiating all previous thinkers.

    **B.** He focused on the power of the individual to use reason to discover truth.

    **C.** He introduced the method of analysis into science and philosophy.

    **D.** He championed the idea that there is a scientific method.

    **E.** He, like his contemporaries who revived ancient atomic theory, created a science based solely on matter and motion, evicting magic and spirit from the study of nature.

    **F.** Last but not least, he (and independently, Pierre de Fermat) invented analytic geometry.

        **1.** Analytic geometry allows us to use the powerful tool of symbolic algebra to solve problems in geometry.

        **2.** We still use Descartes' notation, $x$ and $y$, for the principal unknowns in analytic geometry.

**II.** We focus on Descartes' *Discourse on Method.*

    **A.** We will cut to the chase first: Descartes was greatly impressed with the model of demonstrative science and extolled its power.

    **B.** He wrote, "Those long chains of reasoning, so simple and easy, which enabled the geometricians to reach the most difficult demonstrations, made me wonder whether *all things knowable to man* might not fall into a similar logical sequence" (italics added).

    **C.** If so, he continued, if we make sure our first principles are correct and do our logic correctly, "there cannot be *any propositions* so abstruse that we cannot prove them, nor so obscure that we cannot discover them" (italics added).

**III.** Descartes tells of searching for truth in all the standard ways—in philosophical works, in school, through reasoning, through experience and observation, and through travel—but says he found only widespread disagreement.

    **A.** Skeptics have often done what Descartes did, and he owes a lot to the argument of Michel de Montaigne about how one cannot find truth in the sciences—but Descartes was no skeptic.

    **B.** Instead, Descartes performed a famous thought experiment.
        **1.** He tried to doubt everything: all that he had been taught; the possibly deceptive evidence of his senses; anything involving his sometimes fallible reasoning; even whether he was awake.
        **2.** But he was unable to doubt that he, as a doubting—that is, thinking—being *existed*, saying that it was necessarily true that he, who was thinking that everything was false, "was something."

    **C.** He expressed the "truth" he discovered (psychologically? existentially?) as "I think, therefore I am."
        **1.** We make some psychological remarks about his argument.
        **2.** Existentialists stop here, with this truth; Descartes went on.

    **D.** He now had one truth; first he used it to see how we decide something is true, and then he used it to see if he could deduce other truths from it.
        **1.** He said that he only knew "I think, therefore I am" because it struck him clearly and distinctly, and so he introduced the idea of "clear and distinct ideas" as guaranteed to be true.
        **2.** It is the soul, not the body, whose existence he had proved.

    **3.** He related his truth to proofs of the existence of God, though they had already been formulated by medieval thinkers.

    **4.** Such proofs, he said, show that "it is at least as certain that God, who is this perfect Being, exists, as any theorem of geometry could possibly be."

**E.** In making his method for finding truth in the sciences explicit, Descartes laid down 4 rules:

    **1.** Never accept anything as true unless you "recognize it to be *evidently such*," unless it presented itself "so *clearly and distinctly* to the mind" that it could not be doubted (italics added).

    **2.** Divide each of the difficulties encountered into as many parts as possible, as required for an easier solution; we will call this "divide and conquer" and trace its influence from Adam Smith's division of labor to modern computer science.

    **3.** Always think in an orderly fashion, beginning with the things that are simplest and easiest to understand and gradually reaching more complex knowledge.

    **4.** Make enumerations so complete, and reviews so general, that we can be certain that nothing is omitted.

**F.** That these 4 rules sound a lot like Aristotle or Euclid is no accident; it is right after stating them that Descartes says that "all things knowable to man" might fall into the logical structure of a demonstrative science.

**IV.** Descartes also contributed much to science.

**A.** He argued that space was indefinitely large, filled with matter, and the same in every direction. From this, decades before Newton's *Principia*, Descartes deduced what we now call Newton's first law of motion.

**B.** The motion of vortices (whirlpools, or eddies) in the ether that filled space explained, according to Descartes, why satellites circled the planets and the planets went around the Sun—a highly influential explanation until Newton's theory of universal gravitation replaced it.

**C.** Descartes also made major contributions to the physics of light.

**V.** Descartes' method greatly reinforced the prestige of the demonstrative science model for seeking, as well as presenting and proving, truths that were certain.

**Essential Reading:**

Descartes, *Discourse on Method.*

**Suggested Reading:**

Descartes, *Meditations on First Philosophy.*

Gaukroger, *Descartes.*

Grabiner, "Descartes and Problem Solving."

Popkin, *The History of Skepticism.*

Rubin, *Silencing the Demon's Advocate.*

Russell, *A History of Western Philosophy*, 557–568.

**Questions to Consider:**

1. Go into a dark, quiet room and try to doubt all that you know and the existence of everything. Can you reconstruct for yourself Descartes' experience—that is, prove your own existence to yourself?

2. It has been said that a central tenet of Descartes' philosophy is that the universe is "transparent to reason." Other thinkers believe that the world is mysterious, unknowable even in principle; yet others believe that people should not even try to inquire into how things really are. Where do the thinkers we have seen in the course so far stand on these questions? What do you think?

# Lecture Twenty-Two—Transcript
## Descartes—Method and Mathematics

Welcome back. Here's René Descartes. Descartes is often called the "founder of modern philosophy." People who say that have a lot of good reasons. Here are some of them. Descartes begins by repudiating all previous thinkers, and he did this, not to argue that nothing can ever be known, but to present his own philosophy as superior to everything ever done. That is modern. Descartes focused on the power of the individual to use reason to discover truth. Descartes pioneered the method of analysis (I'll talk a little later about what that is) both for science and for philosophy.

Descartes—along with his contemporary, Francis Bacon, whom we've already met—championed the idea that there is what we now call a "scientific method." Where Bacon championed induction though, Descartes championed deduction. Unlike Bacon, Descartes made major contributions to science. Descartes claimed that his discoveries were the result of using his method.

Also, like several others in the 17th century—including Galileo, who helped revive the Greek atomic theory—Descartes set up a science based solely on particles of matter in motion. It's a mechanistic approach, and that mechanistic approach ultimately banished ideas like magic, spirit, nature abhorring a vacuum, and things like that. The mechanistic approach ultimately banished those ideas from the study of nature.

Last, but certainly not least, Descartes, and, independently, Pierre de Fermat, invented analytic geometry. Analytic geometry lets us use the powerful tool of symbolic algebra to solve problems in geometry. We still use Descartes's notation, $x$ and $y$, for the principal unknowns in analytic geometry—and that's one measure of his great influence.

In this lecture, I want to focus on Descartes's philosophy. More particularly, I want to focus on his best-known philosophical work, the *Discourse on Method*, or, more fully, the *Discourse on the Method of Rightly Conducting the Reason to Find the Truth in the Sciences*. Why is the *Discourse on Method* so important for this course? It's because Descartes's philosophy and his method owes so much to the fact that he was just blown away by the power of the idea of demonstrative science, and the deductive method of ancient geometry.

We will start right with the heart of the matter. Descartes wrote this: "Those long chains of reasoning, so simple and easy, which enabled the geometricians to reach the most difficult demonstrations, made me wonder whether all things knowable to man might not fall into a similar logical sequence." "All things knowable to man." If so, Descartes goes on: "If we make sure our first principles are right, and we do the logical argument correctly," he says, "there cannot be any propositions so abstruse that we cannot prove them, nor so obscure that we cannot discover them." That is claiming a lot!

How did Descartes come to formulate this ambitious program: to imitate the methods of geometric reasoning in order to find all possible truths and prove them? In the *Discourse on Method*, Descartes tells us a story. He tells us about his education—tells us how he read the works of all the philosophers and all the theologians, but how he was never satisfied. He tells how he traveled, and how he reasoned, and how he talked to people. But in all his travels, and discussions, and reading, what he always found was dispute and disagreement—never agreed-upon truth. So Descartes decided that he wasn't going to find the truth in the works of others. He decided that he should look within himself.

Descartes was not the first person to try to do this. In fact, about 80 years earlier, the French philosopher Michel de Montaigne had come to similar conclusions. Montaigne looked at all the writings of past and present, too—and Montaigne decided that nobody had the truth. He didn't claim that knowledge was impossible. Montaigne was not a complete skeptic. In fact, Montaigne says nobody can be a complete skeptic. Montaigne says that even the followers of the ancient Greek skeptic Pyrrho—Pyrrho denied that the evidence of the senses can tell you anything—Montaigne says that even the followers of Pyrrho get out of the road when they see a cart coming.

But Montaigne didn't think that the sciences were ever going to get you any truth. Montaigne even says that the work of Copernicus (that's the guy all of us moderns see as the one who starts the Scientific Revolution), the work of Copernicus, Montaigne says, shows that there's no progress in science. See, in the ancient world, says Montaigne, there were a few people who said that the earth goes around the sun.

Then, he says, the followers of Aristotle and Ptolemy all said that the sun goes around the earth. Now the pendulum has swung back again, and Copernicus has the earth going around the sun again. It's just all like a series of changes in fashion. So, Montaigne says there is no progress in science.

Descartes accepts Montaigne's criticism of the philosophy and science of the past. But Descartes draws different conclusions. Descartes says: "Sure there hasn't been any progress so far, but that's because people have been going about things in the wrong way." It's like what Bacon said, too, you know.

Descartes uses the skeptical criticism of the past, not to despair that we'll never gain knowledge, but instead to construct a new method that will lead to future progress. "If we conduct our reason according to the right method," according to Descartes, "we will indeed find truth in the sciences."

Let's go on with Descartes's personal story and see how he did this. Descartes did a famous experiment in thought. He said he decided, for just one time, that he would completely reject as absolutely false anything of which there could be the least doubt. He wanted to see, if he tried to doubt everything, whether there would be anything left. For instance, since the senses sometimes deceive us, Descartes decided that he would reject all the conclusions of sense experience—and the senses do deceive us. Here's a modern example that I personally happen to like. There are no black dots. But you see them, don't you? Descartes doesn't know about this one, but he does know about illusions of sight. Furthermore, there are also illusions of the other senses. Some people think illusions of sight, you can correct them by touch, but here is an illusion of touch that Descartes gives in his book on optics: "Cross your fingers," he says, "and put a small ball between them. Don't look at it. Just do it by touch."

It will feel as though you are touching 2 little balls. (It works with a pencil, too. You go like this. It feels like you're touching 2 pencils.) The illusion of feeling 2 objects happens because your fingers aren't used to being in this arrangement, and if the outside of these 2 fingers both feel something in their normal arrangement, you perceive that there are 2 of those objects. So, the senses can deceive us—and, for this experiment that Descartes is making in doubting everything that can be possibly doubted, Descartes decides he's going to reject all the perceptions of the senses.

Also, Descartes says: "People make errors in reasoning." He says: "They make such errors even on the simplest topics in geometry." He says: "I can't rely on reasoning either." He says he won't accept any proofs, even proofs that he used to consider valid. Also, Descartes goes on: "Often you can't tell if you're asleep or awake." So maybe everything that he has ever had in his mind was nothing more than the illusion of a dream. So he says he's even going to doubt that he's awake.

All right, don't believe your senses; don't believe any reasoning; don't even believe that you're awake. Is there anything left? Here is what Descartes says: "While I thus wished to think everything false, it was necessarily true that I who thought so was something." That is, while Descartes was doubting everything, he's still conscious of his own doubting and of his existence as a doubter. So, he says, there is something left. "I think, therefore I am." Not even the craziest of skeptics can doubt that one. That's a truth. "I think, therefore I am."

I like to call that Descartes's proof of the existence of Descartes. In a sense, it's an indirect proof: He tries to assume that he doesn't exist, but he finds that he can't do that. Notice that he has proved that he exists only to himself—not to anybody else.

Here's another version of Descartes's proof. This is due to the American philosopher Morris Raphael Cohen of the City College of New York. It seems that a student came up to him and said: "Professor Cohen, can you prove to me that I exist?" Cohen replied: "Certainly—to whom shall I address the proof?" So the student had to kind of re-live Descartes's experiment. By the way, you could do this, too. Go into a dark room where it's quiet and try to doubt everything—including the existence of the objects in the room, that you're awake, that you can reason—doubt that anything you think, or feel, or see, or smell is real. Then try to doubt your own existence. See if you can do it. One of my students tried this when he was in a psychology experiment that put him into a sensory deprivation chamber. You are kind of floating in warm water, so you just don't even feel your own weight. He said he started having hallucinations, but he still wasn't able to doubt that he existed, that he was the being who was having them.

Descartes's argument seems highly original. It is highly original—and yet, in a sense, it's also not original because it's a version of an argument given earlier by Montaigne when he's trying to refute skepticism. I cite this because I think Montaigne's version helps us understand Descartes's argument a little better. Anyway, here it is. The skeptic says: "I doubt everything! I doubt everything!" "And yet," says Montaigne, "the argument takes him by the throat, and forces him to affirm at least one thing: to affirm that he does doubt."

The Existentialists of the 20th century, like Jean-Paul Sartre stop right here. "I think, therefore I exist" is the only truth they admit that we know for sure. They are among the modern heirs of Descartes. But Descartes is after a lot more than that. So, let's follow him some more. Descartes says that this one truth "I think, therefore I am" is so firm and assured that he can safely accept it as the first principle of the philosophy he's after. But what next? "Well," he asks himself, "what am I?" That is, he's just proved the existence of a being—himself. What is the nature of the being whose existence he has just proved? He can think that the world doesn't exist. He can think that his body doesn't exist. It's just as a thinking being that he knows he exists. "Therefore," he says, "I am a substance whose essential nature is only to think."

The part of Descartes that does the thinking—that doesn't need the body. Call it the mind, or call it the soul. The mind would go on existing even without the body—and that's the mind-body distinction, another thing that Descartes is famous for. The mind-body distinction, as explained by Descartes, has a number of implications for the future of science. First of all, the body, since it's material, for Descartes, is just a machine. Even animals, for Descartes, are basically just machines. Thinking about bodies (human bodies, animals bodies) as machines led $17^{th}$- and $18^{th}$-century biologists to a number of important discoveries. For instance, one of Descartes's followers, one Giovanni Alfonso Borelli studied the mechanical advantage of the lever that the arm is, and analyzed the motions of the shoulder considered as a ball-and-socket joint.

In the $18^{th}$ century, Stephen Hales—by the way, Stephen Hales invented the pneumatic trough that you used to collect gases over water if you ever took introductory chemistry—Stephen Hales not only thought about blood vessels as though they were pipes, he was the first to think quantitatively about the pressure within those pipes. So, Stephen Hales was the first ever to measure blood pressure.

Descartes's ideas about mechanistic analysis of bodies are far from obsolete. When I was an undergraduate in the late 1950s, the standard physiology textbook was a book by Anton J. Carlson called *The Machinery of the Body.* The distinction between mind and body was also a weapon useful for early champions of the equality of women. For instance, the French philosopher Poullain de la Barre, he wrote in 1673: "The mind has no sex"—so, important and influential ideas.

Back to Descartes and his search for truth. All right, he now has one truth: his own existence as a thinker, and he's going to milk this one truth for all that it's worth. "I've got one proposition that is true and certain," he says, "so now I've got an example to study to answer this really important question: What is required of a proposition for it to be true and certain?"

"Well," he says, "there's nothing in 'I think, therefore I am' to assure us that it's true unless," he says, "it's that we see very clearly that to think I must exist."

"So," he decides, "I could accept as a general rule that the things which we conceive very clearly and very distinctly are always true." That's the criterion of "clear and distinct ideas," another thing for which Descartes is famous. So Descartes goes on from here to get more truths. First, he gives some proofs of the existence of God. More modern skeptics have been very annoyed at Descartes for doing this. All the same, Descartes does demonstrate God's existence.

We know what "demonstrate" means for people like Descartes—logically deduce something from self-evident first principles. So here's the first argument for God's existence Descartes gives in The *Discourse on Method*. Descartes says: "I doubt. All right, so I'm a being that doubts, but it would be better to know everything and not to doubt. In fact, it's more perfect to know than to doubt. But I have the idea of knowing, as well as the idea of doubting. So where did I get this idea of something more perfect than myself? It has to be from some nature which was itself more perfect, that had all possible perfections, and that's God."

The idea that you can't get the idea of the perfect if you yourself aren't perfect may not seem clear and distinct to us, but it did to Descartes. I have a parallel example that I owe to 2 of my students who came up with it on a very hot day. "Suppose," they said, "that you have an ice-cold can of soda sitting here."

It couldn't have become ice-cold sitting here in the heat. It could only have become ice-cold by being around something—a refrigerator, maybe, or a block of ice—that was very, very cold. "Likewise," they said, "you can't get the idea of perfection sitting here in the midst of all this imperfection—so you must have gotten it from being in the presence of something that was very, very perfect." So I think that's a really helpful analogy to understand how Descartes's argument works.

The idea that God is a being that has "all possible perfections" leads Descartes to another proof of the existence of God. This proof is essentially the medieval one usually credited to St. Anselm of Canterbury in the 11th century. It's called the "ontological proof" (ontological relating to being). Anselm, by the way, gets no credit from Descartes. But anyway, here is the proof as Descartes presents it. Consider a being that has all possible perfections. Let me stop here and elaborate further on what is going on.

Okay, so wisdom is a perfection—so a being with all possible perfections would be wise, and the more wisdom the better, so that being would be infinitely wise. Goodness is a perfection also, so this being must be infinitely good. Now for the big question: Does this being exist? Well, in Western thought anyway, it's better to exist than not to exist—that's an opinion that goes back to Plato. When Plato explains why the god in his dialogue *Timaeus* decides to make the world, Plato says: "This god is good and not envious and therefore he wants all things to be as much like him as possible." Plato's god exists, so he wants everything possible to exist. All right, if it's better to exist than not to exist, existence is a perfection. So this being that has all possible perfections must also have existence. So it must exist.

Descartes goes on to say: "Existence is included in the idea of a perfect being in the same way that the idea of a triangle contains the idea that its angles add up to 2 right angles, or that the idea of a sphere includes the equidistance of all of its parts from the center." He concludes—and he means this as a very strong statement, indeed—"It is at least as certain that God, who is this perfect Being, exists, as any theorem of geometry could possibly be."

Descartes needs God for his philosophy. God is the one who guarantees that our clear and distinct ideas are really true. Otherwise, we might say: "Oh, that seems clear, and distinct, and self-evident, but maybe I'm being deceived by a demon?" But since God is this perfect being, he's the source of all our ideas, and if we get a clear and distinct idea from God, we know it must be true.

At this point, many modern people get really disappointed with Descartes. People might say: "Oh, he did something really new for a minute, but look, now we're back with the old medieval theology." But we are far from finished with what Descartes has to say in the *Discourse on Method*.

Descartes gave 4 rules for his new method. Here's a shorthand version of them. These are his rules for seeking truth in the sciences. They are also—and this is no accident—the way to do geometry. Let's look at them each in turn. "The first rule," Descartes says, "is never accept anything as true unless you recognize it to be evidently so; include nothing that doesn't present itself so clearly and distinctly to your mind that you can't doubt it."

Here are some examples of my own, drawn from this course. We certainly could accept "Between 2 points we can construct a straight line" or "If equals are added to equals the results are equal" as meeting Descartes's criterion: self-evident, striking us clearly and distinctly. Descartes's first rule certainly seems to reflect Euclidean practice.

"The second rule," Descartes says, "is to divide each of the difficulties which I encounter into as many parts as possible, as might be required for an easier solution." That's the method of analysis. Here are some examples—again, my examples and not his. In geometry, say we wanted to find the area of a circle. First, we see that we can approximate that area by the area of a polygon with lots and lots of sides. Then, we realize that we can treat a polygon as made up of triangles.

So, we first figure out how to find the area of a triangle. See, we've analyzed the circle problem into triangles. Then, we put the triangles together, so we can find the area of any polygon, and then we get the circle by approximating its area by the area of a polygon with lots, and lots, and lots of sides.

This "divide and conquer" approach to solving problems in various fields is alive and well today. Among many examples I could cite from history, consider Adam Smith's analysis (I choose the word carefully) of the economic success of Europe in the $18^{th}$ century: Smith credits it to the "division of labor." To make an industrial product, you divide the task into a large number of subtasks and then carry out each subtask as efficiently as possible. Henry Ford used this same idea—analyzing the large problem (building a car) into small sub-problems as he built his factories to mass-produce cars. Each worker and each machine did just one little part of the task of building a car, but did it with optimal efficiency, and the end product was not only reliable, it was cheap.

If you've ever done any computer programming, you'll recognize Descartes's method: You divide your big problem into little problems, and you write a little sub-program for each little problem—and you put all those sub-programs together into one large program. In fact, when I used to teach computer science, I quoted Descartes's Rule Two at the top of my syllabus, as you see here. "My rule was to divide each of the difficulties which I encountered into as many parts as possible, as might be required for an easier solution." In fact, Descartes even starts the *Discourse on Method* by saying: "Hey, if this seems too long for you to read all at once, read one chapter at a time." Divide and conquer.

Back to the rules. "The third rule," Descartes says, "is to think in an orderly fashion, beginning with the things that were simplest and easiest to understand, and gradually and by degrees reaching toward more complex knowledge." So—again, this is my example—we might start with points and lines, and then go on to triangles and circles, and then go to 3 dimensions—and eventually get to things like there are 5 regular solids and the volume of a sphere. In fact, you would start with "elements"—simple things that more complicated things were made up of—and go on from there.

"The last rule," Descartes says, "is always to make enumerations so complete, and reviews so general, that I would be certain that nothing was omitted." So—my example again—first we could prove that if 2 straight lines come together, the angles add up to 2 right angles; then we could prove that if the angles add up to 2 right angles, this line at the bottom is a straight line, and then we let the lines cross each other, and then we prove that the vertically opposite angles so formed are equal. Now we've got the topic of intersecting lines and their angles completely covered. That Descartes's 4 rules sound a little like Aristotle and a lot like Euclid, that's no accident. It's right after stating these rules that Descartes says what I quoted at the beginning of this lecture.

Now that we know how he got there, it's worth quoting that again:

> Those long chains of reasoning … [which] enabled [the geometers] to reach the most difficult demonstrations, [had] made me wonder whether all things knowable to man might not fall into a similar logical sequence. [He goes on:] If so, we need only refrain from accepting as true that which is not true, and carefully follow the order necessary to deduce each one from the others, and there cannot be any propositions so abstruse that we cannot prove them, not so [recondite] that we cannot discover them.

If we stick to the form of demonstrative science, of Euclidean geometry, we can eventually know everything that can be known. This is sometimes called "Descartes's dream." Wow.

But it doesn't work, right? As Francis Bacon said, and as Newton was to say half a century after Descartes: "Hey, you can't just sit in your study and reason your way to the laws of nature." But Descartes gave it his best shot. There are laws of nature that can be—and were—discovered by Descartes reasoning according to these rules. Here's one example: "Space," says Descartes, "is indefinitely large, and it's the same in every direction. Every point in it is the same." How do we know this, by the way? Well, it strikes us as clear and distinct. There's no reason it should be otherwise. All right, but anyway, accept that.

Now imagine a body in space that is at rest. Why should it move if nothing acts on it? Any point has the exact same nature as any other point, so there's no reason for the body to go elsewhere, to prefer any other point to the one where it's at; so, if nothing acts on a body at rest, it stays where it is. Suppose a body is moving, with nothing pushing on it in any direction. Why should it speed up?

Every point has the same nature, so there's no reason the body should want to get to some point faster, or avoid some point by going slower; there's no reason for it to turn from a straight path to the right or to the left, since every direction is the same.

So a body at rest with nothing acting on it stays in its state of rest, and a body with nothing acting on it that's moving in a straight line at a constant speed continues to do so. Descartes publishes this in 1644 and calls it the first law of nature. In 1687, Newton made it his own first law of motion.

Descartes did a lot more physics—very influential in his time. He was a Copernican (Descartes was), and he tried to give a physical explanation for the structure of the solar system. He said that space was filled with matter— a kind of ethereal matter, which he, indeed, called the "ether." In the ether, bodies of grosser matter produced eddies or whirlpools, which he called "vortices." You know, it happens just the way a rock in a streambed produces eddies in the water around it.

For instance, there's a vortex around the earth. That's why the moon orbits the earth—the moon is caught in this little whirlpool in the ether, just the way little chips of soap circle the drain when you let the water out of the bathtub. Likewise, the vortex around the sun (pictured here) carries the planets around—and that's why the solar system works the way it does, according to Descartes. This was a very influential explanation, until Newton's theory of universal gravitation replaced it. Also, Descartes explained light as vibrations in this physical ether that fills space—and that's an explanation that persisted into the 20th century.

Descartes was also an independent co-discoverer of what is called "Snell's law," for the refraction of light—how light is bent when it goes from air into water or glass. That's why, by the way, we see the straw as bent at the surface, where the water and the glass interface here. Although we don't have time to describe it, Descartes's independent co-invention of analytic geometry was incredibly influential; that's why we call the $x$ and $y$ coordinates I mentioned before Cartesian coordinates.

So Descartes's influence, both on science and on philosophy, has been enormous. Descartes had a large number of followers among scientists, not to mention philosophers. In the next 2 lectures, we'll see some examples of Descartes's philosophical legacy.

But to conclude this lecture, here's what I think is the most important. Descartes's "method" greatly reinforced the prestige of the "demonstrative science" model of reasoning. The model of demonstrative science was, after Descartes, thought of as useful for seeking—as well as presenting and proving—truths that were certain. Thank you.

# Lecture Twenty-Three
# Spinoza and Jefferson

**Scope:** We now give 2 especially striking examples of the way the ideal of demonstrative science, as described by Aristotle, exemplified by Euclid, and reaffirmed by Descartes, influenced important thinkers in the modern world. We first look at Benedict (Baruch) Spinoza's *Ethics Demonstrated in Geometrical Order*. Spinoza built a deductive system in a completely Euclidean form, with precise definitions of the key terms, with axioms governing the meaning and use of the basic concepts, and with the major conclusions expressed as theorems. We will focus on Spinoza's proof that God exists. We then look at the American Declaration of Independence. We will observe how Thomas Jefferson gave the document the form of a Euclidean proof, beginning with axioms (self-evident truths), continuing by adducing evidence for the "if" part of an if-then inference, and concluding with the "therefore" clause that actually declares the colonies' independence from Britain. We will close by considering the power of the logical argument to discover new truths and to drive moral and political action.

# Outline

**I.** We turn to 2 examples of important thinkers who used the model of demonstrative science to reason about questions far from geometry and crucial to human beings and society.

**II.** Benedict (Baruch) Spinoza wrote an *Ethics Demonstrated in Geometrical Order*.

    **A.** Spinoza's goal was to *demonstrate*, in the geometrical sense of the word, the truth about God, nature, and humanity.

    **B.** Spinoza's major conclusions have been influential. Einstein, a kindred spirit, wrote, "I believe in Spinoza's God who reveals himself in the orderly harmony of what exists."

    **C.** The substance of the work uses philosophical and theological ideas from classical Judaism, Aristotelian philosophy, medieval theology, and Descartes' philosophy.

**III.** Spinoza's *Ethics* is structured exactly like Euclid's *Elements*.

    **A.** Part 1, "Of God," begins with definitions of the key terms, then axioms, and then propositions to be proved.

        **1.** For instance, "cause of itself" is defined as "that whose essence involves existence"; "substance" is defined as "that which is in itself and is conceived through itself"; and "God" is defined as "Being absolutely infinite, that is to say, substance consisting of infinite attributes."

        **2.** Among Spinoza's axioms are "from a given determinate cause an effect necessarily follows"; "the knowledge of an effect depends upon and involves the knowledge of the cause"; and "the essence of that thing which can be conceived as not existing does not involve existence."

        **3.** Among Spinoza's propositions is "God ... necessarily exists"; he gives 4 proofs, the first being an indirect proof.

    **B.** We briefly address why Spinoza chose the methods of geometry and ask whether Spinoza's proof merely unpacks his verbal definitions of the basic concepts or whether something else is involved.

**IV.** We now move about 100 years further on, to Thomas Jefferson and the Declaration of Independence.

    **A.** Bypassing the introductory sentence, we focus on the wording of what comes next: "We hold these truths to be *self-evident*" (italics added).

        **1.** Jefferson's first draft, instead, held the truths to be "sacred and undeniable."

        **2.** Benjamin Franklin—a first-rate scientist, discoverer of the theory of electrical charge, and member of the Royal Society of London—is probably the one who suggested the more scientific and, incidentally, more Cartesian term "self-evident."

        **3.** Among these self-evident postulates are that "all men are created equal," that the goal of government is to secure human rights, and "that whenever any Form of Government becomes destructive of these ends, it is the Right of the People to alter or abolish it, and to institute new Government."

    **B.** Jefferson could now postulate, "If King George's government is destructive of these ends, then it is the right of the people to alter or abolish it." This postulate has the form "If $p$, then $q$."

**C.** To show that declaring independence is legitimate, Jefferson must prove the statement $p$; he can then use the valid argument form "If $p$, then $q$. $q$; therefore, $p$."

**D.** The declaration first states that the king of Great Britain has violated the rights of the people by his bad government; it then says, "To *prove* this, let Facts be submitted to a candid world" (italics added).

**E.** The greater part of the declaration is a long list of tyrannical things done by King George that make him "unfit to be the ruler of a free people," a list which is intended to prove $p$.

**F.** Having proved $p$, Jefferson can conclude $q$.
    **1.** The actual statement of the part of the document that declares independence makes clear he has this geometric-style argument structure in mind.
    **2.** The last paragraph of the Declaration of Independence begins: WE, **THEREFORE**, the REPRESENTATIVES of the UNITED STATES OF AMERICA … solemnly publish and declare, That these United Colonies are, and of Right ought to be FREE AND INDEPENDENT STATES" (bold-face added to "therefore"; otherwise, typography as in the official printed version).

**G.** We briefly discuss why Jefferson chose to cast the declaration in the form of a geometric proof.

**V.** The form of the declaration, as well as its content, has been very influential.

**A.** It has encouraged drawing logical conclusions from the statement "All men are created equal"; we give some examples.
    **1.** The Seneca Falls declaration asserting the rights of women begins, "We hold these truths to be self-evident; that all men and women are created equal."
    **2.** We quote also from speeches by Frederick Douglass, Martin Luther King Jr., and Harvey Milk.

**B.** Other countries, after 1776, declared their independence; we look at some examples and see how they followed the logic of the American declaration.

**C.** We conclude by restating the idea that understanding the logical implications of general principles is a powerful force, both to discover new truths and to marshal principles that can drive political action.

## Essential Reading:

The Declaration of Independence.

Nadler, "Baruch Spinoza."

Russell, *A History of Western Philosophy*, 569–580.

## Suggested Reading:

Armitage, *The Declaration of Independence*.

Becker, *The Declaration of Independence*.

Cohen, *Science and the Founding Fathers*.

Curley, *Behind the Geometrical Method*, 3–50.

Spinoza, *Ethics Demonstrated in Geometrical Order*.

Wolfson, *The Philosophy of Spinoza*, chaps. 1–6.

## Questions to Consider:

1. Spinoza's conception of God got him into a lot of trouble, including accusations of atheism. What other views of God are common in Jewish or Christian thought, and why might those holding such views have attacked Spinoza's approach?

2. We have all read many political arguments justifying various courses of action; few have the form of a Euclidean theorem. How else might Jefferson have tried to convince his contemporaries of the justification of the colonies' declaring independence? Why are arguments like his not common today?

# Lecture Twenty-Three—Transcript
## Spinoza and Jefferson

Welcome back. Today, we're going to look at 2 important thinkers who used the deductive model of demonstrative science to reason about questions very far from geometry—questions of major importance to human beings and to society. Each of these 2 people was well educated in mathematics and science, and each actually contributed to the science of his day. They were Benedict Spinoza and Thomas Jefferson.

The work of Spinoza that we will concentrate on shows its relevance to our course in its title: *Ethics Demonstrated in Geometrical Order*. For Jefferson, we'll look at the Declaration of Independence of the United States of America.

Let me first introduce Spinoza. Spinoza was brought up in the Jewish community in Amsterdam, in Holland, a haven of religious freedom in 17th-century Europe. But Spinoza was a critic of all traditional religion, to put it mildly. Spinoza was influenced by Descartes, but went much farther; Spinoza was the consummate rationalist. He was educated in the sciences—he actually worked as an optician. But he had larger goals than mere science. Spinoza's goal was to demonstrate—in the geometric sense of "demonstrate"—the truth about God, nature, and humanity.

For Spinoza, religion ought to be limited only to what is absolutely essential for an ethical life: that there is a Supreme God who loves justice and charity. "Otherwise," Sponoza says, "we should abandon the superstitions that pass as religion, we should abandon the belief in miracles—a belief that just reveals our ignorance of the universal law of cause and effect. Instead, we should live a life governed by reason." A modern kindred spirit to Spinoza was Albert Einstein, who wrote: "I believe in Spinoza's God, who reveals himself in the orderly harmony of what exists."

In his book, *Ethics Demonstrated in Geometrical Order*, Spinoza drew on ideas from medieval philosophy and from the philosophy of Descartes. But his *Ethics* is structured exactly, but exactly, like Euclid's *Elements*. We'll be looking at Part One of the *Ethics*, the part about God. In that part, I'm going to concentrate on the proposition in which Spinoza proves the existence of God. Let me say at the outset that Spinoza's argument is abstract, difficult, and full of technical terminology. But what's important for our purposes is to keep our eyes on the form.

Spinoza's *Ethics* begins with definitions of the key terms that he will use. Just as I did when we looked at Euclid, I will give only the definitions most relevant to the proposition that we're interested in—that God exists.

So here's Spinoza's Definition 1: "By 'cause of itself,'" Spinoza says, "I understand that whose essence involves existence, or that whose nature cannot be conceived unless existing." That language ought to remind you a little bit of Descartes's version of the ontological proof of the existence of God—a resemblance that is, of course, no accident.

Next, substance. "By 'substance,'" Spinoza writes, "I understand that which is in itself and is conceived through itself; in other words, the conception of which does not need the conception of another thing from which it must be formed." Maybe some examples will help here. The concept of "substance" goes back to Aristotle; it's found in medieval philosophy; it's found in Descartes. For Descartes, for instance, both mind and matter are substances. For Descartes, a substance has one principal attribute that constitutes its nature or its essence.

For Descartes, as I guess we've seen, the principal attribute of mind is thought. That's not surprising. In the same way, the principal attribute for matter, according to Descartes, is that it occupies space.

Want to know what an "attribute" is? Spinoza defines that also: "By 'attribute,'" Spinoza says, "I understand that which the intellect perceives of substance as constituting its essence." That's consistent with the examples I gave about mind and matter.

Now "essence" is another already-existing philosophical concept. Spinoza doesn't define it here. But the "essence" of something is that which makes it what it really is—again, its essential nature. Let me ask a question that can serve as an example: What's the essence of being a human being? Being male? No. Female? No. Tall? No. Short? No. Being of a particular ethnicity? None of those is the essence of being human. Maybe Aristotle got it right, the essence of a human being, when he said: "Man is a rational animal."

In any case, here is one more definition from Spinoza: "By God," he says, "I understand Being absolutely infinite, that is to say, substance consisting of infinite attributes, each one of which expresses eternal and infinite essence."

I kind of wonder whether Aristotle would approve of these definitions as meeting his criterion of defining things in terms of other things that are better known and prior. Some of the terms that Spinoza uses in his definitions are already known to students of medieval philosophy and students of Descartes. So, in that respect, Spinoza is doing what Euclid did and what Aristotle said—you may think differently.

Now for some of Spinoza's axioms. These are supposed to be self-evident, clear, and distinct assumptions. First axiom (I'm using his numbers again), the first one I've got here anyway. Axiom [2]: "That which cannot be conceived through another must be conceived through itself." Axiom [3]: "From a given determinate cause an effect necessarily follows; and, on the other hand, if no determinate cause be given it is impossible that an effect can follow." [Axiom 4]: "The knowledge of an effect depends upon and involves the knowledge of the cause." These last 2 may remind us of Aristotle's views about the relationship between causality and explanation.

One more. [Axiom 7]: "The essence of that thing which can be conceived as not existing does not involve existence." Let me do that one again. "The essence [the essential nature] of that thing which can be conceived as not existing does not involve existence."

I'm not going to prove all of Spinoza's propositions, but I do need to state one of them—the one he needs in his proof of the existence of God. So this is one of his propositions. [Proposition 7]: "It pertains to the nature of substance to exist." I'm not going to do the proof, but I will say that his proof of this proposition draws on the idea of substance being a cause of itself, and thus on his first definition.

Now we're ready for the big one, his Proposition 11 as it happens: "God, or substance consisting of infinite attributes, each one of which expresses eternal and infinite essence, necessarily exists."

Here are his proofs of the existence of God. It's pretty important, so he gives several proofs. The first is sort of an indirect proof. It depends on "substance with all attributes" having to have existence. It will remind you, of course, of Descartes's first proof of the existence of God. Spinoza says: "If the existence of God be denied, conceive, if it be possible, that God does not exist." Then it follows from Spinoza's Axiom 7 that God's essence does not involve existence. But that, in the light of his Proposition 7, that it pertains to the nature of substance to exist is absurd. He says: "Therefore, God necessarily exists." I don't know if you're totally convinced by that one—but, anyway, there it is.

Spinoza's next proof rests on his idea of the universality of cause and effect. He starts by saying: "For the existence or non-existence of everything there must be a reason or cause." The cause for something's existence or non-existence can be within the thing's nature, or it can be outside of its nature. In case the reader doesn't understand—I think that's a good bet, by the way—Spinoza gives an example of a cause from within the thing's nature and one outside of it: Both examples come from geometry.

First, a cause from within its nature for non-existence: "A square circle can't exist because then its nature would involve a contradiction." That can't happen. Second, from outside the thing's nature: "Whether a particular triangle or a particular circle exists or not," Spinoza says, "depends on the order of physical nature." "In either case, though," he says, "if there is no cause or reason that hinders a thing from existing, it exists necessarily."

Let's apply all of this to the existence of God. If there's no reason or cause to stop God from existing, God must exist. Why isn't there such a cause to stop God from existing? Suppose there were. If there is a cause that stops God from existing, and that cause is outside of God's nature, that cause would have to be just as powerful as we claim God to be. But then that would be God. If the cause that stops God's existing was from within God's nature, then there would be a contradiction in God's nature. But to say that there's a contradiction in the nature of a Being that is absolutely infinite and perfect is absurd.

Spinoza finishes by saying: "There's no existence that we can be more sure of than of the existence of an absolutely infinite or perfect Being—that is to say, God." It's interesting to compare the way Spinoza uses reason to argue for God with the way his contemporary Pascal used it. Pascal—as you'll recall—used probability, and he said that although we can't prove God's existence by reason, reason does make us conclude that God is the best bet. But Spinoza thought that you could prove the existence of God by the same methods that we use in geometry. Of course, my goal in this lecture is not to convince you that there's a God, especially a God who is substance with infinite attributes. My goal is to give another example of the impact of geometry on Western thought. Spinoza's case shows how a serious and influential philosopher used the methods of geometry, the deductive form of a demonstrative science, to construct a theological and ethical system that would be certain and true.

But I do want to ask one question about how convincing Spinoza's arguments might be. In a way, it seems to me, anyway, as though the conclusion, the existence of God as Spinoza defined God, it's deliberately packed into the definitions and into the axioms. To me, anyway, his arguments sound like the argument: "All golden mountains are golden," or, after defining "bachelor" as "unmarried man," like an argument that "All bachelors are unmarried." Spinoza's arguments are more complex than the examples I just gave, but they seem to me, anyway, to be of the same kind.

Spinoza and his contemporaries may have been able to consider his axioms as self-evident, since the terms "essence," "existence," "substance," "attribute," "cause," and "effect" were all staples of the prevailing systems of philosophy. But I still think there's a sense in which Spinoza's proofs proceed just by analyzing the terms that he uses. But, you know, Euclid's geometry does not seem to be like that. Reading the postulates and definitions, I appeal to you, yes, I think we don't feel that you can just analyze the terms and come up with the Pythagorean Theorem or "the exterior angle of a triangle is bigger than either opposite interior angle." We certainly need Euclid's diagrams to prove any of his theorems. I'm not going to pursue this point right now, but it's going to be really important later on when we look at the philosophy of Immanuel Kant. So I at least wanted to get those ideas out there on the table.

Last and I think the most interesting question about Spinoza for us: Why does he present his conclusions about religion—and ethics—in what he calls "geometrical order"? Here I think you can anticipate the answer. This, the deductive model of demonstrative science, is the way you get truth and certainty.

Plato and Aristotle are cheering you on. It's a way of cashing in on the prestige of the Euclidean model—and it's an endorsement of Descartes's ideal: The rational method, the mathematical method, is the road to truth. Many other people in Spinoza's time felt the same way. So it was, for instance, that in 1687 Isaac Newton presented his theory of the universe, beginning with 3 assumptions that he called "axioms" or "laws of motion." These are the laws that appear in every first-year physics book. In fact, Newton wrote his great book, the *Mathematical Principles of Natural Philosophy* in exactly the Euclidean form—with definitions, axioms, and propositions.

Now we move ahead about 100 years farther on, to 1776 and the Declaration of Independence of the United States. You all know how the main section starts, right? We hold these truths to be self-evident; that all right angles are equal. No, that's not what it says, but that's a little bit what it sounds like, isn't it?

By the way, the author of the Declaration of Independence, Thomas Jefferson, knew more of the mathematics of his time than any president that the United States has ever had. Jefferson actually contributed to the mathematics of voting theory, and also to designing ploughs using calculus whose shape minimized the resistance produced as the plough moved through the soil. Like Spinoza, Jefferson used ideas from earlier thinkers—in Jefferson's case, from John Locke and from various Scottish and Swiss political philosophers. But, also like Spinoza, Jefferson put the earlier ideas to novel use and cast them into a new—and in Jefferson's case—incredibly influential form.

Let us look now at the Declaration of Independence, first at the correct wording of the sentence I mangled before: "We hold these truths to be self-evident, that all men are created equal." Self-evident. Jefferson's first draft didn't say that. Jefferson's first draft instead said: "We hold these truths to be sacred and undeniable." A leading historian of the Declaration, Carl Becker said that the change from "sacred and undeniable" to "self-evident" was suggested by Benjamin Franklin. The change makes the statement "We hold these truths to be self-evident, that all men are created equal" sound more scientific.

Benjamin Franklin wasn't just a kite-flyer. He was a first-rate scientist. He devised the first quantitative theory of electricity, including inventing the concept of electric charge. So he's a likely source. But in any case, that term "self-evident" carries with it all the authority of the scientific and mathematical views stemming from Euclid and Descartes.

Jefferson is clearly presenting, at the beginning of the Declaration, a set of postulates. Among these self-evident postulates are: "that all men are created equal"; "that the goal of government is to secure human rights"; "that human beings set up government"; and: "Whenever any Form of Government becomes destructive of these ends, it is the Right of the People to alter or abolish it, and Institute new Government." That postulate logically implies this statement: "If King George's government is destructive of these ends, then it is the right of this people to alter or abolish it and set up a new government." That self-evident truth has the form "If $p$, then $q$."

To show that declaring independence from Britain is legitimate, what Jefferson needs to do is prove this statement $p$: "King George's government doesn't secure human rights; it is destructive of the purpose of government." If Jefferson can prove $p$, then he can invoke this valid argument form: If $p$, then $q$. $p$. Therefore, $q$. So that he could conclude $q$: that the people of the colonies have the right to abolish King George's government and set up a government of their own.

This is exactly the way the Declaration of Independence is constructed: First, a postulate that if government is destructive of the ends of securing rights, it is the right of the people to abolish it and replace it. Second, Jefferson states that the King of Great Britain has violated the rights of the people of the American colonies by his bad government. Third, Jefferson then undertakes to prove this, to prove $p$. I'm not the one who's saying "prove." The Declaration of Independence explicitly says: "To prove this, let Facts be submitted to a candid world."

"To prove this …" What facts are to be submitted to a candid world? The longest part of the Declaration of Independence is a long list of tyrannical things done by King George that make him, as the Declaration says, "unfit to be the ruler of a free people"—a list whose purpose is to prove the statement I've called $p$: King George's government is destructive of the ends of government.

Once he has proved $p$, Jefferson can then conclude $q$. If $p$, then $q$. $p$. Therefore, $q$. That's what he does in exactly those words. The statement of the operative part of the document—that actually declares independence—makes clear that he wants us to recognize this as a geometric-style argument.

So I'm going to show you on the screen the actual Declaration of Independence of the United States, the key statement in the type-face of the original printed version (except that there's one word that's in bold-face, and I put it in bold-face):

> WE, **THEREFORE**, the REPRESENTATIVES of the UNITED STATES OF AMERICA … solemnly publish and declare, That these United Colonies are, and of Right ought to be FREE AND INDEPENDENT STATES."

So the "therefore," the word that signals in geometry that we've proved something, is part of the official founding statement of the United States of America.

Jefferson himself did not always live up to the principles stated here. He owned slaves, both men and women, and may have had a sexual relationship with one of the women. But if we choose to criticize him, we do so in terms of the general principles that he himself set forth. We use logic to argue that he was wrong.

Why did Jefferson write the Declaration of Independence in the form of a geometric argument? Why didn't he say something like this: "You guys are pushing us around. You English changed your own government when you set up a parliamentary democracy under William and Mary. Your king has no God-given right to rule us. So we're going to do the same thing, set up our own government. We hereby declare our independence. Goodbye." Why didn't he do that?

I think it's because Jefferson, like Spinoza, knows and respects mathematics—and he assumed that his audience, both in America and in Britain, shared those views. Jefferson wanted to gain, for his argument, all the prestige carried by mathematical arguments.

Euclidean geometry, as we'll see in even more detail in the next lecture, was the 18th-century ideal of reasoning. The Declaration of Independence is going to be an argument? Then, it should be the strongest possible type of argument. So it has turned out to be. A simple statement that the colonies were rebelling against Britain because the British were oppressing them might have been enough for the immediate purpose of the revolution that began in 1776. But then others would not have looked at the Declaration and said: "Hey, if all human beings are created equal and entitled to human rights, how about us?" Because of the way Jefferson cast the Declaration, people have said that. Many people have found, in the logical implications of the ideas in the Declaration of Independence, arguments and inspiration for their own fights for freedom. Here are some examples.

In 1848, 5 women meeting in Seneca Falls, New York, adopted a Declaration of Sentiments and Resolutions urging the United States to grant women the right to vote. The Seneca Falls Declaration marks the beginning of the organized women's movement in the United States. The main author of this document was Elizabeth Cady Stanton. She was also an Abolitionist, by the way. She carefully followed the structure of the Declaration of Independence. Just reading the beginning makes this point clear: "We hold these truths to be self-evident: that all men and women are created equal."

So the promise of universal human rights logically implied the inclusion of women. Reminding Americans that they had signed on to these principles, and thus to the logical consequences of these general principles, had great rhetorical force.

In 1852, in a speech given at Rochester, New York, in honor of the Fourth of July, the anniversary of the Declaration of Independence, of course, the ex-slave and Abolitionist Frederick Douglass said on behalf of all African-Americans: "The signers of the Declaration of Independence were … great men. … They were statesmen, patriots and heroes, and for the good they did, and the principles they contended for, I will unite with you to honor their memory." But, Douglass went on: "Are the great principles of political freedom and of natural justice, embodied in that Declaration of Independence, extended to us? … Would to God, both for your sake and ours, that an affirmative answer could be truthfully returned!" Thus the ex-slave, Frederick Douglass, in 1852.

Here's another example, from Martin Luther King's "I Have a Dream" speech in 1963. Dr. King said:

> When the architects of our republic wrote the magnificent words of the Constitution and the Declaration of Independence, they were signing a promissory note to which every American was to fall heir. This note was the promise that all men, yes, black men as well as white men, would be guaranteed the unalienable rights of life, liberty, and the pursuit of happiness.

Harvey Milk, the first openly gay man elected to a major public office, also found inspiration in these ideas, saying: "On the Declaration it is written, 'All men are created equal.' No matter how hard you try, you can never erase those words. That is what America is!" It's not just for Americans. Other peoples have declared independence, and many other Declarations have been written. Many national movements for independence have stressed their commitment to universal human rights, rather than just their own nation's desire to be rid of a particular set of rulers. They don't always use the Euclidean form for their argument, but they do use phrases that echo the generality of Jefferson's words.

For instance, from the Venezuelan Declaration of Independence, 1811:

> We meet considering the full and absolute possession of our Rights … We cannot, nor ought not, preserve the bonds that hitherto kept us united to the Government of Spain: … like all the other nations of the world, we are free, … authorized … to take amongst the powers of the earth the place of equality which the Supreme Being and Nature assign to us.

From the Declaration of the Texas Republic in 1836:

When a government has ceased to protect the lives, liberty, and property of the people, from whom its legitimate powers are derived, ... the inherent and inalienable right of the people to appeal to first principles, ... enjoins it as a right ... to abolish such government, and create another in its stead.

This one ends: "We therefore, the delegates of the people of Texas ... do hereby resolve and declare, that our political connection with the Mexican nation has forever ended, and that the people of Texas do now constitute a free, sovereign, and independent republic."

The next one is from Liberia, 1847. It says:

We recognize in all men, certain natural and inalienable rights: among these are life, liberty, and the right to acquire, possess, enjoy and defend property. ... The right therefore to institute government, and to all the powers necessary to conduct it; it's an inalienable right, and cannot be resisted without the grossest injustice.

It, too, ends with a "therefore": "Therefore in the name of humanity, and virtue and religion." They appeal to the world to recognize and respect their independence as a nation.

Perhaps surprisingly, the Declaration of Independence of Viet Nam, in 1945, follows Jefferson's pattern. It begins: "All men are created equal. They are endowed by their Creator with certain inalienable rights, among these are Life, Liberty, and the pursuit of Happiness." As with the American Declaration, they list a long list of bad things that they say their French rulers have done, and the Vietnamese document closes similarly also: "For these reasons, we, members of the Provisional Government of the Democratic Republic of Viet Nam, solemnly declare to the world that Viet Nam has the right to be a free and independent country."

Last, here is Czechoslovakia in 1918, declaring their independence from the Hapsburg Empire:

We reject the sacrilegious assertion that the power of the Hapsburg and Hohenzollern dynasties is of divine origin. The Hapsburgs broke the contract with our nation by illegally hampering our rights ... and we therefore refuse longer to remain a part of Austria-Hungary.... We make this declaration on the basis of our historic and natural right.

Arguments of this form—you believe in these general ideas; therefore, it logically follows that they should be applied to our case—such arguments have tremendous persuasive power.

In the next lecture, we'll look further at some of the leading ideas of the 18<sup>th</sup> century and see what they owe to the prestige, structure, and truth claims of Euclidean geometry. To conclude this lecture, though, let me take off from what Dr. King said: "The promise of the ideas in the American Declaration of Independence is powerful. The generality of the principles of the Declaration—the fact that they say that they apply to the whole human race—is part of what gives them their power."

Spinoza, and Jefferson, and their intellectual heirs all agree that working out the logical implications of general principles is a force not only to discover new truths, but also to drive moral and political action. Thank you.

# Lecture Twenty-Four
## Consensus and Optimism in the 18<sup>th</sup> Century

**Scope:** In the 18th century, mathematics in general and geometry in particular served as sources of inspiration—and as models—for philosophers hoping to achieve the same kind of certainty in their own field as geometers had achieved in theirs. We saw this with Thomas Jefferson; now we focus on Voltaire, Isaac Newton, Gottfried Wilhelm Leibniz, Edward Gibbon, and the Marquis de Condorcet. Voltaire, among others, argued that what made the conclusions of geometry so certain was that nobody disputed them. We explain Voltaire's version of this argument and describe its historical antecedents, but we will focus especially on the role of Euclidean geometry in justifying Voltaire's views. Moreover, in the 18th century, algebra also became an example of the progress of mathematics. We will see how this progress, together with Descartes' philosophy, helped buttress the characteristic optimism of many 18th-century philosophers—concluding with the soaring predictions by the Marquis de Condorcet for the perpetual progress of the human race.

## Outline

I.  As we have seen, the prestige of mathematics in general and geometry in particular has been used to promote arguments outside of mathematics that have the same form.

   **A.** Besides using the demonstrative science model of argument, geometry had other properties that philosophers thought they could emulate.

   **B.** One such property was that there seemed to be universal agreement about geometry.

   **C.** Another property was the method of analysis, as championed by Descartes.

   **D.** The success of algebraic symbolism, as exemplified by analytic geometry, gave all of mathematics much greater power.

   **E.** As mathematics became more successful, it served as the model for the social and political progress people wanted to achieve.

**II.** Since the Renaissance and Reformation, philosophers had been concerned with what has been called the "the problem of the criterion."

    **A.** Since there were many competing candidates for "the truth," be it in religion or philosophy or politics, it was natural to ask, by what criterion can one identify the truth?

    **B.** Since geometry seemed to have achieved truth, perhaps seeing how this had been done would help solve the problem of the criterion.

    **C.** Descartes' "clear and distinct ideas" were one possibility, though even Descartes conceded that sometimes it was hard to tell what we conceive clearly and distinctly.

    **D.** Many 18th-century thinkers, most eloquently Voltaire, said that a criterion of truth, clearly visible in the case of geometry, is that everybody agrees about it.

**III.** Descartes's method of analysis proved very influential in the 18th century.

    **A.** Adam Smith applied analysis when examining the division of labor in *The Wealth of Nations*.

    **B.** Gaspard François de Prony employed this division of labor when he was tasked with creating a set of logarithmic and trigonometric tables for the French Revolutionary government.

    **C.** Charles Babbage also used the concept in developing a computing machine. He wrote about the process in an essay called "On the Division of Mental Labour."

**IV.** After the invention of analytic geometry, the power of symbolic algebra was added to the success stories about mathematics.

    **A.** Symbolism made mathematics more general and helped to discover new truths.

        **1.** Isaac Newton called algebra, because of its power of generalization, the "universal arithmetic."

        **2.** Symbolic algebra is a powerful instrument of discovery because it reveals the process behind the answer.

    **B.** Gottfried Wilhelm Leibniz invented a notation for his calculus that is still prized for its problem-solving power.

        **1.** Leibniz also envisioned a more general symbolic language that would allow us to reduce all problems to algebraic calculation, making him a forerunner of modern symbolic logic.

**2.** Attempts were made in other fields to devise symbolism with the same generalizing and problem-solving properties; the best known is in chemistry, where chemical symbols and balancing chemical equations were called a chemical algebra.

**V.** The methods of mathematics and science were billed as the keys to human progress.

    **A.** The mathematically based science of Isaac Newton was, in the 18th century, widely thought of as demonstrating the superiority of modern thought over all that had come before.

        **1.** Newton was buried in Westminster Abbey, as if, noted Voltaire, he had been a king.

        **2.** Alexander Pope wrote, "Nature and Nature's laws lay hid in night; God said, 'Let Newton be!' And all was light."

    **B.** The ideas of Pierre-Simon Laplace and many 18th-century thinkers were built upon the belief that there was one set of scientific laws; that these showed what all the diverse things in the world had in common; and that human beings, by means of reason, could find these laws.

    **C.** The Scottish philosopher Francis Hutcheson applied mathematical methods to society, famously concluding that the best action is the one that produces the greatest happiness for the greatest number.

**VI.** Many 18th-century thinkers thought that the progress of mathematics, science, and technology presaged perpetual and universal progress for the entire human race.

    **A.** For instance, Edward Gibbon, in his *Decline and Fall of the Roman Empire*, wrote that European civilization would not collapse the way Rome did, because of the "progress in the arts and sciences" and because any "barbarian" challenger to Europe, in order to become strong enough to surpass Europe, would need to develop enough to "cease to be barbarous."

    **B.** The Marquis de Condorcet epitomizes the promoters of the idea of progress.

        **1.** Condorcet linked the existence of algebraic symbolism with Descartes' goal of using the mathematical method to find out "all things knowable to man," saying that algebra "contains within it the principles of a universal instrument, applicable to all combinations of ideas" that "could make the progress of every subject embraced by human intelligence as sure as that of mathematics."

**2.** Condorcet concludes his work on the future progress of the human mind by saying, "the progress of the mathematical and physical sciences reveals an immense horizon ... a revolution in the destinies of the human race."

**VII.** The new contributions to mathematics of the period since the Renaissance and the exciting possibilities of scientific progress that these generated drove the Enlightenment idea of progress, fed the American and French revolutions, and still animate thought and action today.

**Essential Reading:**

Gay, *The Enlightenment*.

**Suggested Reading:**

De Condorcet, *Sketch for a Historical Picture*.

Lovejoy, *The Great Chain of Being*, chaps. 6–7.

Popkin, *The History of Skepticism*.

Redman, *The Portable Voltaire*.

**Questions to Consider:**

**1.** "If it is true, then everyone agrees" does not logically imply "If everyone agrees, then it is true." Surely Voltaire and other partisans of the consensus approach to truth knew that much logic. Why, then, did they find this theory so attractive?

**2.** Why has the idea that science will continue to progress, and therefore that society will also necessarily progress, fallen out of favor in the modern world?

# Lecture Twenty-Four—Transcript
## Consensus and Optimism in the 18th Century

Hello again. Without much exaggeration, I could have called this lecture "Mathematics is What Made 18th-Century Western Thought." We've just seen how the prestige of mathematics in general and of geometry in particular promoted arguments in other areas that had the form of geometric proofs. But mathematics had other properties that philosophers thought that they could emulate. Here are 4 of them: universal agreement, the method of analysis, algebraic symbolism, and the idea of progress.

One property (the first one, then) is that in mathematics, unlike in any other area of thought, in mathematics there seem to be universal agreement about what is true. Another property was the method of analysis, as championed by Descartes. Analysis—taking things apart and studying the whole by means of studying the individual parts—that seemed to be a key to understanding both nature and society. Also, the success of algebraic symbolism—exemplified by analytic geometry—gave all of mathematics much greater power.

Finally, mathematics, and the science based on it, was becoming more and more successful—making it a model for the social and political progress that people wanted to achieve. The 18th century was the heyday of these views: universal agreement, analysis, algebraic-style symbolism, and progress—all modeled on the success of mathematics. So, let us look together at each of these in turn.

First, universal agreement—outside of mathematics, it seemed to 18th-century thinkers as though everybody was always arguing. Especially since the Renaissance and the Reformation, philosophers had been concerned with what has been called "the problem of the criterion." See, the Renaissance revived the views of Plato, who disagreed with the views of Aristotle that had been around so prestigiously in the Middle Ages. Renaissance humanism argued for different values than did Christianity. The Reformation meant that Protestants disagreed with the Catholic Church over what true Christianity was. The new astronomy of Copernicus competed with the old system of Ptolemy, and the new material philosophy of atomism competed with the older physics of Aristotle. So, since there were so many competing candidates for "the truth," it was natural to ask:

What is the criterion by which we could distinguish what is true from what is false?

That's the problem: the criterion. Since geometry seemed actually to have achieved truth, maybe seeing how geometry had done this would help solve the problem of the criterion. Descartes had proposed one possible solution to the problem of the criterion—clear and distinct ideas—but Descartes himself conceded that sometimes it was hard to tell what we're conceiving clearly and distinctly. Some philosophers denied that there were any ideas innate to the mind that had the properties Descartes wanted them to have— John Locke, for example disagreed. So what's a philosopher to do?

Many 18th-century thinkers had a different candidate for the criterion for truth—a criterion clearly visible in the case of the truths of geometry: Everybody agrees about them. This view was most eloquently expressed by Voltaire, and he is worth quoting at some length. "The universal agreement about mathematics," says Voltaire, "stands in sharp contrast to the case of religion, where there are so many competing sects. Every sect, of every kind, is a rallying point for doubt and error," writes Voltaire, and then he gives a long list of philosophical and religious sects. "But by contrast," he says, "there are no sects in geometry: One does not say, 'I'm a Euclidean, I'm an Archimedean.'"

"When the truth is evident," Voltaire says, "it's impossible for parties and factions to arise." Voltaire goes so far as to say that in England nobody says: "I'm a Newtonian" because those who have read and understand Newton can't possibly refuse their assent to the truths that Newton teaches.

Voltaire gives another interesting example: A man named Rymer has gathered all the historical records of the Tower of London, but there aren't any Rymerians—he's just the guy who found the truth.

After those positive examples of agreement, Voltaire looks at religion. Look, the Hindus say you can't eat beef. The Jews and the Muslims say: "Beef is fine—you just can't eat pork." They disagree. According to the criterion laid down by Voltaire, that means they're all wrong. "And what," Voltaire asks, "is the true religion? It's the one everybody agrees on." Is there such a thing? Yes—Voltaire says: "All religions agree on this: There is a God, and you must be just. This, where all religions agree, this is true; the things on which they differ are false."

Voltaire goes on to say: "Somebody who sees that the square of the hypotenuse of a right-angled triangle is equal to the sum of the squares of the other two sides is not of the sect of Pythagoras." "Likewise," says Voltaire, "when you say that blood circulates, you aren't of the sect of William Harvey, who discovered it. In these cases, you merely agree with the truth as demonstrated by Pythagoras or Harvey, and" says Voltaire in exactly these words, "the whole world will always be of your opinion."

So here's Voltaire's answer to the problem of the criterion. In the concluding words of this little essay on sects, Voltaire says: "This is the character of truth: It is for all time, it is for all men, it has only to show itself to be recognized, and one cannot argue against it." When there are disputes—as there are over morality, for instance—the true answer is that which all the different sides have in common. In his little essay on morality, Voltaire wrote: "There is but one morality, as there is but one geometry."

My students get justifiably indignant at the idea that what people all agree on has to be the truth. After all, they say, it used to be universally accepted that slavery was part of the human condition and that men were superior to women. You could also argue that there's a logical fallacy involved here. To say: "If something is true, then it's universally accepted" does not mean that "If something is universally accepted, then it's true." But, like it or not, this criterion of universal agreement was widely accepted in the 18$^{th}$ century.

People pushed that criterion into intercultural and international issues. When questions about human nature arose and people wanted to distinguish human nature from what was merely custom, 18$^{th}$-century thinkers were inclined to say: "What differs from society to society is custom; what all societies have in common is human nature." As in mathematics, so in the world at large—universal agreement is a mark of truth.

Now to the 18$^{th}$-century society's second inheritance from mathematics: Descartes's method of analysis. I've already mentioned Adam Smith. As I said when I was talking about Descartes, Smith, in his *Wealth of Nations*— published, by the way, in the same year as the Declaration of Independence, in 1776—Adam Smith analyzed the competitive success of economic systems by means of the concept of division of labor. The separate elements of the economy, each acting as efficiently as possible, provided for the overall success of the manufacturing process. In fact, you can see Smith's method of analysis in the most famous quotation from Adam Smith's *Wealth of Nations*. Smith said that every individual in the whole economy, while striving to increase his individual advantages, is "led as if by an invisible hand to promote ends which were not part of his original intention"—that is, as Smith analyzes what would be the best economic outcome for the whole of society, he finds that it comes from individuals—each maximizing their own situations.

As you might expect, the French especially liked Descartes's method of analysis, as refracted through Adam Smith as well. Here's a fascinating example. After the French Revolution, the government needed a set of logarithmic and trigonometric tables; these are useful for navigation and for gunnery. A man called Gaspard François de Prony had the job of calculating these tables, but he didn't do it on his own. Prony decided to do it by applying Adam Smith's ideas about the division of labor. So first, Prony analyzed the job of calculating these mathematical tables into its components, and then he organized a group of people into a hierarchical system to do the various parts of the job. Here's how he planned it.

A few mathematicians decided which functions to use. Then, a larger group of technicians reduced the job of calculating the functions to a set of simple additions and subtractions of pre-assigned numbers. Finally, a large number of low-level human "calculators" carried out the actual additions and subtractions. The idea of dividing up a piece of intellectual work like this seemed very exciting to a young Englishman called Charles Babbage. Charles Babbage was, in the early 19th century, a pioneer of the digital computer, because he took Prony's analysis of computation and embodied the instructions and the resulting calculations in a machine. The machine, by the way, was never quite completed—but here is a piece of it that's now in the Science Museum in London.

Babbage wrote about what he was doing in an essay called "On the Division of Mental Labour." It's an amazing triumph of the method of analysis, although technical limitations prevented Babbage's design from having any immediate practical success. He's got the idea of the digital computer and the program; it's the same hierarchical "divide up the job into little pieces."

Talking about Babbage takes us also to the next of the mathematical topics I was addressing: symbolic algebra. Babbage was philosophically a follower of Leibniz, in that both men championed the power of symbolic notation. Symbolic notation, in the view of many mathematicians, could make mathematical calculation almost mechanical—so that a mechanical computer was a natural idea. Actually Leibniz had designed the calculator himself—not a computer, but a calculator. But there are more profound consequences of symbolism in algebra. Symbolism both made mathematics more general and helped discover new truths.

Let's see a little how that works. First, making mathematics more general—Newton called algebra "universal arithmetic." Let me tell a story that will make this point clear. When I was in the third grade, we had to learn what my teacher called the "100 addition facts"; you put them up on the blackboard, and we had to learn them. They ranged from $0 + 0 = 0$ up to $9 + 9 = 18$. Okay, so I'm looking at them all, and I noticed that $9 + 7$ was 16 and that $7 + 9$ was also 16.

That these sums were the same was new to me, and I didn't know how to ask the question right. I should have asked: "Is addition commutative?" But what I said was: "$9 + 7$ is 16 and $7 + 9$ is 16. Is it always like that?" and my teacher said: "Yes, $9 + 7$ is always 16." But what I wanted to know wasn't the answer to $9 + 7$, but whether it really never mattered in what order you add numbers—because that would mean that there aren't 100 addition facts to memorize; there are only about half that many.

But we can save even more time than that. Algebra is a universal arithmetic. So it lets us write down an infinite number of addition facts—things like $9 + 7 = 7 + 9$, $2 + 10 = 10 + 2$, and so on—all at once. Watch me, I'm going to write down infinitely many addition facts on the screen. $a + b = b + a$. For any 2 numbers represented by the symbols $a$ and $b$, $a + b = b + a$. Okay, that's generality.

Second, general symbolic notation can be an instrument of discovery. For instance, suppose there's a problem and someone tells you that the answer is 16. You don't have any idea how they got that answer. Maybe they got it by adding 9 and 7. But maybe they got it by dividing 64 by 4, or multiplying 8 times 2. But we know exactly where $a + b$ comes from; it comes from adding $a$ and $b$. As Viète, the inventor of general symbolic notation in algebra, liked to say: "In algebra the operations leave their footsteps behind."

Let me show you a deeper example than $a + b$. Suppose we have a quadratic equation. You remember quadratic equations. Look here: $6x^2 - 19x + 15 = 0$, and someone tells you: "Hey, the solutions are $1\frac{2}{3}$ and $1\frac{1}{2}$." That's nice to know, but it doesn't give you any help at all if you need to solve a different quadratic equation. But if you use general symbolic notation, this is every quadratic equation: $ax^2 + bx + c = 0$, $a$ not equal to 0 because if $a$ was 0 it wouldn't be a quadratic equation. If you solve this one—by using the standard algebraic technique called "completing the square"—the details don't really matter, but if you solve this one, what you do is you find the solution in terms of any 3 numbers—$a$, $b$, and $c$. Solutions being $(-b + \sqrt{b^2 - 4ac})/2a$ and $(-b - \sqrt{b^2 - 4ac})/2a$.

That's the solution. The operations we performed to get this solution have left their footsteps behind. So now we have the solution for every quadratic equation in terms of the numbers in it. If you've solved this one quadratic equation, you've solved them all. For instance, the solution to my first example is just that formula where $a = 6$, $b = -19$, and $c = 15$. So symbolic algebra is a wonderful instrument of discovery.

This problem-solving power of algebra—whether applied to geometry by Descartes, or to practical problems in banking, or to probability theory—this problem-solving power inspired mathematicians to apply symbolism to other areas. For instance, Leibniz (1646–1716)—Leibniz invented a symbolic notation for the calculus that we still use today because it has problem-solving power. That isn't all that Leibniz wanted to do. Leibniz also envisioned a more general symbolic language—a symbolic language for all of thought, a symbolic language that would allow us to reduce all questions, in philosophy and politics as well as science, to algebraic calculation. This makes Leibniz a pioneer of what is now called "symbolic logic." Leibniz imagined that if we had such a general symbolic language, there would be no more disputes in the world. If you and I were to disagree, one of us would say to the other: "Let us calculate!"—and we would work out the consequences of our initial assumptions algebraically, and we would get an unambiguous answer.

Scientists in other fields in the 18$^{th}$ century also tried to devise symbolism with the same generalizing and problem-solving properties. Take chemistry. If you've ever taken a chemistry course and learned to balance chemical equations, you know how helpful that chemical notation is. Chemical symbols, like $H_2O$ and $CO_2$, and the process of balancing equations using them, were called by the great 18$^{th}$-century chemist and discoverer of oxygen, Lavoisier a "chemical algebra." So Descartes and Fermat, their analytic geometry was just the first of a whole series of successful applications of symbolic notation to achieve scientific progress.

Speaking of progress, we now come to the 18$^{th}$-century idea of progress in general. Descartes's introduction of a rational method for finding truth about everything, not just mathematics, was linked in many people's minds with the actual success of mathematics and science.

Humanity finally had found a way to move forward. Here's a typical 18$^{th}$-century quotation that links the method of universal agreement with the idea of progress. It's from Colin MacLaurin, the Scottish mathematician that I work on, and when I read you this, notice the idea of progress, the idea of right method, the idea of universal agreement, and the rejection of sectarian arguments—all wrapped up in the same quotation from Colin MacLaurin.

He writes: "We are now arrived at the happy era of experimental philosophy; when men, having got into the right path, prosecuted useful knowledge. The arts received daily improvements ... societies of men, with united zeal, ingenuity and industry, prosecuted their enquiries into the secrets of nature, devoted to no sect or system."

The mathematically based science of Newton was, in the 18th century, widely thought of as demonstrating the superiority of modern thought over everything that had come before. Voltaire found a dramatic way to point this out by saying Newton was buried in Westminster Abbey, just as if he had been a king. The poet Alexander Pope, in the 18th century, combining biblical and Platonic metaphors, wrote these lines: "Nature and Nature's laws lay hid in night; / God said, 'Let Newton be!' And all was light."

As we already saw when we looked at the ideas of Laplace, many 18th-century thinkers believed that there was one set of scientific laws, that these showed what all the diverse things in the world had in common—and that human beings, by means of reason as Descartes had promised, could find these laws.

God made the universe according to law, and when we think something is due to chance or isn't good, we're just wrong—these misperceptions are just the results of our lack of knowledge.

Alexander Pope put this into poetry. Let's quote him again:

> All nature is but art unknown to thee;
> All chance, direction which thou canst not see;
> All discord, harmony not understood;
> All partial evil, universal good.

That's human progress—finding the perfect order in the universe.

Many people also thought that mathematical methods could produce a science of society. For instance, the Scottish philosopher Francis Hutcheson wrote an essay called "On Computing the Morality of Actions." Hutcheson famously said that the best action is the one that produces the "greatest happiness for the greatest number." That "let's maximize good things" idea of Hutcheson's, which is linked to Descartes's method of analysis, continued to bear fruit as classical economics developed—from Adam Smith onward. So not only was there going to be progress in the sciences, there was going to be social progress as well.

In fact, many 18th-century thinkers thought that the progress of mathematics, science, and technology guaranteed perpetual and universal progress for the whole human race. In the 17th century, both Francis Bacon and Descartes—each with their different scientific methods—had already predicted this. They had each explained that the relative lack of progress in the past [was because of] the lack of the right way of proceeding, and both Bacon and Descartes predicted a glorious future for the human race once everybody proceeded by the right method.

In the 18th century, scientists had personally experienced substantial scientific progress, and they were quite ready to ascribe it to having the right method, including the use of mathematics. Eighteenth-century scientists built their own accounts of what they were doing—and requests for support from both business and government—on the views and the rhetoric of Bacon and Descartes.

There really weren't a lot of major technological changes in the 18th century due to cutting-edge science; there were a few—Benjamin Franklin's electrical theory, for instance, produced the lightning rod. The analytical method in general, as we saw in discussing Adam Smith, certainly improved agriculture, manufacturing, navigation, and so on. But wait! Perpetual human progress? What about the unquestioned historical fact that all the great empires have declined? Had the 18th century somehow figured out how to avoid this fate? Yes, many people thought that they had figured this out. We're going to listen to Edward Gibbon.

Gibbon was the author of the famous work, *The Decline and Fall of the Roman Empire*—another product, by the way, of the year 1776. "Was European civilization," he asked himself, "so brilliant at the end of the 18th century, doomed to collapse the way Rome had?" Gibbon's style is much more elegant than mine—so let me read how Gibbon asked the question: "We may inquire with anxious curiosity, whether Europe is still threatened with a repetition of those calamities that formerly oppressed the arms and institutions of Rome." Gibbon's answer to that question was "No." What he says was informed by his mastery of Roman history, of course—but, more important, permeated with the optimism of the 18th century.

Here's what Gibbon writes: "The experience of four thousand years should enlarge our hopes, and diminish our apprehensions; we cannot determine to what height the human species may aspire in their advances towards perfection; but it may safely be presumed that no people, unless the face of nature is changed, will relapse into their original barbarism."

But Mr. Gibbon, I ask: "Isn't that what happened to Rome—that they were conquered, they were conquered by armed invading barbarians?" Gibbon's reply is essentially: "Yes, but never again." How come? Gibbon writes: "Mathematics, chymistry [sic], mechanics, architecture, have been applied to the service of war; and the adverse parties oppose to each other the most elaborate modes of attack and defence … An industrious people should be protected by these arts." Here's the crucial point—Gibbon goes on: "Europe is secure from any future irruption of Barbarians; since, before they can conquer, they must cease to be barbarous." See, you can't score a victory over European civilization, Gibbon is saying, unless you are civilized enough to have a sophisticated science and technology. Spears and arrows won't do it. Using that 18th-century phrase: "the arts and the sciences," where we would say "science and technology," Gibbon says that: "The arts and sciences have been successively propagated; they can never be lost."

Gibbon ends his great book on the decline and fall of the Roman Empire by saying: "We may therefore acquiesce in the pleasing conclusion that every age of the world has increased, and still increases, the real wealth, the happiness, the knowledge, and perhaps the virtue, of the human race." So mathematics and science will give us universal progress.

It's hard to beat Gibbon, but there's somebody who can. Our last prophet of progress is the Marquis de Condorcet. Condorcet was a mathematician. He contributed both to the calculus and to probability theory. He was also a pioneer in the application of mathematical methods to the study of society, especially in the theory of voting. But he's most famous for writing a book called *A Sketch for a Historical Picture of the Progress of the Human Mind.*

Condorcet, like Voltaire and Jefferson, was an advocate for human reason, and for the ability of reason and science to promote freedom and progress. To get an idea of his politics, here's what he says he's for—"an understanding of the natural rights of man, the belief that these rights are inalienable and [cannot be forfeited], liberty of thought and letters, of trade and industry, the alleviation of the people's suffering, the [elimination] of all penal laws against religious dissenters and the abolition of torture and barbarous punishments, and legislation in conformance with reason and nature."

Condorcet saw science in general, and mathematics in particular, as the keys to progress. He linked the existence of algebraic symbolism with

Descartes's goal of using the mathematical method to find out "all things knowable to man." Condorcet agreed with Leibniz about algebraic symbolism's universal power.

Condorcet said that algebra "contains within it the principles of a universal instrument, applicable to all combinations of ideas." An instrument, Condorcet continued that could "make the progress of 'every subject embraced by human intelligence … as sure as that of mathematics.'" Let me repeat that: the right symbolism could "make the progress of 'every subject embraced by human intelligence … as sure as that of mathematics.'"

Condorcet concluded his book on the progress of the human mind with these words: "The progress of the mathematical and physical sciences reveals an immense horizon … a revolution in the destinies of the human race."

Condorcet is a hard act to follow, but let me try, anyway, to conclude this lecture—actually to conclude this whole section of the course, since Lectures Thirteen through Twenty-Four have all been about the triumphal march of geometry and its methods, from ancient Greece to 1800.

In conclusion, then: The new contributions to mathematics between the Renaissance and 1800, and the exciting possibilities of the scientific progress that these generated, reinforced older ideas based on the success of Greek geometry and geometry's claims to truth.

All of this helped to drive the Enlightenment idea of progress, fed the American and French Revolutions, and still animate thought and action today. The leading spirits of the 18th century saw Euclidean geometry as uniquely and necessarily true and as a model for all human thought.

Geometry, along with the rest of mathematics and with science, was to be the engine of perpetual progress—intellectual, material, and moral progress—for the whole human race. Thank you.

# Lecture Twenty-Five
# Euclid—Parallels, Without Postulate 5

**Scope:** We now return to Euclid's geometry, as we prepare for the philosophical confrontation between the Euclidean and non-Euclidean versions of space. In the next 2 lectures, we study the propositions and the proofs in Euclid's theory of parallel lines, Propositions 27–32. We will see how much more careful Euclid is than the average high school textbook, and we will emphasize the philosophical, as well as mathematical, importance of the careful logical structure he provides. In the present lecture, we will go through the proofs of 3 of these theorems that do not logically require Euclid's famous parallel postulate. We conclude by discussing what it means to say that Euclid proved that parallel lines actually exist.

# Outline

I.  We return now to geometry, looking at a crucial part of Book 1 of Euclid's *Elements*: the theory of parallels, especially Propositions 27–32.

   A.  This section, more than any other, has been lauded as epitomizing the logical soundness of Euclid's reasoning.

   B.  But the theory of parallels also provoked a 2000-year-long challenge to Euclid's demonstrative science.

II. We briefly discuss Euclid's Postulate 5, repeating the statement of the postulate, which we saw in Lecture Sixteen.

   A.  Euclid's fifth postulate provides exactly—no more, no less—what he needs to prove key propositions about parallel lines and their properties.

   B.  Yet it is not a postulate of which one could say, "We hold this truth to be self-evident."

   C.  We mention some criticisms of this postulate, going back to antiquity.

**D.** There is a common substitute postulate, which is the logical equivalent of Euclid's postulate: Given a line and a point not on that line, there is only one line parallel to the given line through that point. This is often called Playfair's axiom, after John Playfair, who used it in his influential 18th-century geometry textbook.

**E.** It was not until the 19th century that mathematicians realized, after many failed attempts, that Postulate 5 could not be proven from Euclid's other assumptions.

**III.** What is logically wonderful about Euclid's Propositions 27–32 is how he first carefully proves the basic things that can be proved *without* Postulate 5, calling upon that postulate—which he uses explicitly exactly once—only when it is absolutely necessary to do so.

    **A.** We state Propositions 27 and 28, designating the angles by number for easy identification.

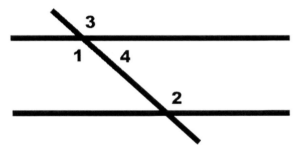

        **1.** Proposition 27: If 2 lines are cut by a third line such that the alternate interior angles (angle 1 and angle 2) are equal, then the lines are parallel.

        **2.** Proposition 28: If 2 lines are cut by a third line such that the corresponding angles (angle 2 and angle 3) are equal, or so that the interior angles on the same side of the third line add up to 2 right angles (angle 2 + angle 4 = 2 right angles), then the lines are parallel.

    **B.** We note that these statements all have the form "If something is true of the angles, then the lines are parallel."

    **C.** We emphasize that each such statement does not logically imply its converse, whose form would be "If the lines are parallel, then something is true of the angles." If the converse turns out to be true, it will require a separate proof.

**IV.** We now prove Proposition 27.

    **A.** We begin by recalling the definition of parallel lines: lines in the same plane that never meet.

    **B.** We ask, "How will we be able to prove, with no previous propositions about parallel lines, that 2 lines are parallel, since that would mean showing that these 2 lines, though infinitely extended, never intersect?"

    **C.** We can do this only by assuming that the lines meet and showing that this leads to a contradiction.

    **D.** Thus this has to be an indirect proof, and we carefully exhibit its logical form.

    **E.** The proof starts with the given fact that angle 1 = angle 2. If the lines meet, though, by Proposition 16, angle 1 must be greater than angle 2—a contradiction, so the lines cannot meet.

    **F.** This proof does not require the fifth postulate, but it does require Proposition 16, that the exterior angle of a triangle is greater than either opposite interior angle.

        **1.** Euclid's proof of Proposition 16, we recall, required Postulate 2, that a line can be extended to any desired length.

        **2.** This fact will become important later on when we discuss non-Euclidean geometry.

**V.** We now prove Proposition 28.

    **A.** The additional results required are Propositions 13 and 15, which were discussed in Lecture Twenty-One.

        **1.** Proposition 15, which we need first, is that when 2 lines intersect, the vertically opposite angles are equal (thus angle 3 = angle 1), and we can use Proposition 27 to show the lines are parallel.

        **2.** Proposition 13 states that when a line is set up on another line, the 2 angles formed add up to 2 right angles. Thus angle 2 + angle 4 = angle 1 + angle 4, and subtraction of angle 4 yields angle 1 = angle 2, so we can use Proposition 27 again.

    **B.** We note that, since the proof of Proposition 28 depends on the proof of Proposition 27, the proof of Proposition 28 also depends on the truth of Postulate 2.

**VI.** As Plato observed, geometry is hypothetical, and Euclid's theorems are true only if the postulates are.

   **A.** As Aristotle, Descartes, Spinoza, and many others would prescribe, if you want truth and certainty, construct a demonstrative science based on self-evident first principles.

   **B.** As we look at all the apparently parallel lines in our immediate environment—the lines on the pages of a book, the sides of books on bookshelves, the sides of floorboards or of hexagonal tiles on a floor—we see the truth of Propositions 27 and 28 and can use them to construct these parallel lines.

**VII.** We now briefly address Proposition 31, the construction of a line through point A parallel to BC.

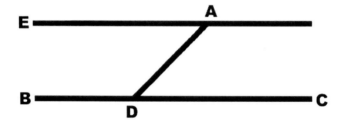

   **A.** We go through Euclid's proof, which requires being able to copy an angle, something Euclid proved we can do in his Proposition 23.

   **B.** We draw any line from A to the line BC, hitting it at D; we copy the angle ADC as EAD, making these equal alternate interior angles, which means the new line EA is parallel to the original line BC.

   **C.** We observe that Proposition 31 otherwise depends only on Proposition 27, so the construction of parallel lines does not presuppose Postulate 5, although it does presuppose Postulate 2.

   **D.** So Proposition 31 proves that parallel lines exist.

      **1.** Before he proved the existence of parallel lines, Euclid used Proposition 30 to show that parallel lines, if they exist, are unique. This proposition had to rely on a result whose proof required Postulate 5—namely, the result of Proposition 29, which will be covered in the next lecture.

**2.** This leaves us with a question: Since there are parallel lines, what properties must they have?

### Essential Reading:

Heath, *The Thirteen Books of Euclid's Elements*, vol. 1, Book 1, Propositions 27–28, 31.

### Suggested Reading:

Heath, *The Thirteen Books of Euclid's Elements*, vol. 1, notes to these propositions, 308–311, 316.

Katz, *A History of Mathematics*, chap. 3 (chap. 2 in the brief edition).

### Questions to Consider:

**1.** If Postulate 2 is false on a surface, does that mean that there cannot be any parallel lines on that surface?

**2.** What is your response to the criticism that Euclid is only using lots of complicated logical machinery to "prove" something that is obvious from looking at a diagram?

# Lecture Twenty-Five—Transcript
# Euclid—Parallels, Without Postulate 5

Welcome back. In the last lecture, I said that Lectures Thirteen through Twenty-Four revealed the triumphal march of Euclidean geometry. In the next 9 lectures, Lectures Twenty-Five through Thirty-Three, I'm going to blow the semi-religious worship of Euclid's geometry out of the water.

In order to do this, I first need to take you into the heart of the Euclid worship: Euclid's logically superb theory of parallel lines. Then, we'll see various philosophers and scientists argue that this theory was necessarily true, and that it was part of the fabric both of all human thought and of the universe. Then, we will re-live the shocking invention of non-Euclidean geometries. Finally, we'll look at the impact of those new geometries on life and thought. The present lecture, and the immediately next lecture, will be almost entirely about proofs in the geometry of parallel lines. I promise you, though, that the work you put in to understand them in detail will pay off handsomely in the lectures to come.

The basic propositions of Euclid's theory of parallels are found in Book 1 of the *Elements*, Propositions 27 through 32. This section of the *Elements* has long been lauded as epitomizing the logical soundness of Euclid's "long chains of reasoning reaching the most difficult demonstrations."

But this same theory of parallels also provoked a 2000-year-long challenge to Euclid's ideal of demonstrative science. The bone of contention, as we saw in Lecture Sixteen, was Euclid's fifth postulate: "If a straight line falling on 2 straight lines makes the interior angles on the same side less than 2 right angles, then the 2 straight lines eventually meet on that side." That is, in this picture, if angle A plus angle B add up to less than 2 right angles, those 2 sideways lines have to meet on that side.

The fifth postulate provides exactly—no more, no less—what Euclid needs so he can prove some—but not all—of the key propositions about parallel lines and their properties. Yet Postulate 5, as I've already observed, is not one of which we could say: "We hold this truth to be self-evident." This postulate was criticized already in antiquity—for instance by a leading commentator on Euclid, a guy called Proclus.

Here's what Proclus had to say about Postulate 5: "Postulate 5 ought to be struck from the list of postulates altogether because it's a theorem and requires for its demonstration a number of definitions as well as theorems." He thought it was a theorem, so Proclus tried to prove Postulate 5 from the rest of the postulates, but—as various people soon pointed out—Proclus's proof introduced yet more assumptions. There were other attempts to prove Euclid's fifth postulate in the Greek-speaking world, but none of them was accepted.

After the fall of Rome and the Islamic conquests, mathematicians in the Islamic world, beginning with Greek geometry, advanced mathematics considerably—as our use of the Arabic world "algebra" for the theory and practice of solving equations can serve to illustrate.

A number of prominent mathematicians in the Islamic world tackled the problem of proving Postulate 5 from Euclid's other postulates. Specialists in science in the Islamic world would recognize all of their names, but I'm just going to mention one of them right now. We'll get back to them later. The one I'm going to mention is Omar Khayyam; [he] also was a famous poet. Attempts in the Islamic world helped to clarify which results were logically equivalent to Euclid's fifth postulate, and the work in the Islamic world influenced later European attempts to prove that postulate. But, for our present purposes the thing to notice is that in the Islamic world, as in the Greek, nobody succeeded in proving Euclid's fifth postulate from his other postulates.

Okay, I said logical equivalents to Postulate 5. One of these "logical equivalents" to Postulate 5 is worth special mention. It's the postulate about parallel lines commonly used in modern geometry textbooks. It says: "Given a line and a point not on that line, there is only one line parallel to the given line through that given point." Call that the "uniqueness of parallels."

Proclus proved that this—the uniqueness of parallels—was logically equivalent to Postulate 5. His reward for doing this is that it got named after somebody else. It got named after a man of the 18th century, John Playfair—incidentally John Playfair was the brother of William Playfair, who invented the bar graph and the pie chart.

John Playfair wrote an influential geometry textbook, and it used as its parallel postulate this "only one parallel" property, so that property got called "Playfair's axiom." So people who had previously been trying to prove Euclid's fifth postulate, some of them tried to prove Playfair's axiom, as well as trying to prove Euclid's original version, and they kept trying to do this until the 19th century, when mathematicians finally realized—with far-reaching implications—that proving Euclid's fifth postulate from Euclid's other assumptions was impossible.

I mention this 2200 years of history—and the 2200-year survival of Euclid's fifth postulate—to motivate our respect for Euclid's theory of parallels. As we'll see in a few minutes, Euclid first carefully proves all the basic things that he can prove without Postulate 5. Euclid calls on that postulate—which he uses explicitly exactly once—he calls on it only when it is absolutely necessary for him to do so.

In this lecture, we'll look together at the basic propositions about parallel lines that Euclid proves without any appeal to Postulate 5. I'll do the ones that do need Postulate 5 in the next lecture. First, I'll just state the 3 theorems that do not need Postulate 5 that are in this little piece of the *Elements*. Here is the relevant picture for all of them.

Proposition 27 states that: "If 2 lines are cut by a third line such that the alternate interior angles are equal [that is, in this picture angle 1 and angle 2 are equal], then the lines are parallel."

Proposition 28 states 2 facts: "If 2 lines are cut by a third line such that the corresponding angles [that's angle 2 and angle 3 in my picture] are equal, then the lines are parallel. Last: "If 2 lines are cut by a third line such that the interior angles on the same side [angle 2 and angle 4 in my picture] such that angle 2 plus angle 4 add up to 2 right angles, then the lines are parallel." Notice that all of these statements have this form.

If something is true of the angles, then the lines are parallel. As I emphasized when we were talking about "if-then" statements in the lecture on logic, Lecture Eighteen, such a statement does not logically imply its converse; the form of the converse would be: "If the lines are parallel, then these things are true of the angles." If the converses both turn out to be true, each one is going to require its own distinct proof.

Finally, here is the last of the propositions in this section of the *Elements* that does not require the fifth postulate. It's Proposition 31: "Through a given point, we can construct a straight line parallel to a given straight line." What I'm going to do first is to prove Proposition 27, that if what we call the alternate interior angles are equal, then the lines are parallel—that is if angle 1 is equal to angle 2—see, they're interior to the parallel lines; that's why they're called interior, and they alternate from this side of the third line to this side of the third line, so they're called alternate interior angles. "If the alternate interior angles (angle 1 and angle 2) are equal, then the lines are parallel." That is Proposition 27.

Now we've got to start the proof (since we want to prove the lines are parallel) by recalling the definition of parallel lines: Those are "lines in the same plane that never meet." But it seems like we've got a problem. This is the very first proposition about parallel lines. We know nothing else about parallel lines, except for the definition.

So, how in the world can we prove that 2 lines are parallel? This would mean going off to infinity in one direction to show that they don't meet, and even if we could do that—which we can't—then we'd have to go off to infinity again in the other direction to show they don't meet. It seems impossible—but we're in luck. Maybe we don't know much about parallel lines yet, but we do know a lot about logic. We have a powerful logical tool: the process of proof by contradiction, or indirect proof.

So what we're going to do, then, as we follow Euclid, is we're going to assume that the lines aren't parallel—that they actually meet. We already know a lot of geometry about lines that meet.

Hopefully, then, assuming that the lines meet will lead us to a contradiction. If we get a contradiction, then that will mean that the assumption that the lines meet has to be false. If they can't meet, we'll have proved that they must be parallel.

So, let's see how Euclid does this. The proposition says that "If angle 1 is equal to angle 2, then the lines are parallel." So the "if" part—we're given that; we're given that angle 1 is equal to angle 2. That's true. For the sake of argument, let's assume that those lines aren't parallel—they meet. So, the picture doesn't look like this, but looks like this; the lines meet.

If the lines meet, we've got 3 lines here that form a triangle. We know a lot of things about triangles. But in particular, we know a proposition from Lecture Twenty-One (you may remember that as Proposition 16)—we know a proposition that says that: "In any triangle, the exterior angle is greater than either opposite interior angle." Therefore, angle 1, which is an exterior angle for this triangle, has got to be greater than angle 2, which is one of the opposite interior angles.

But wait. We're given that angle 1 is equal to angle 2. These angles can't both be equal to each other and have angle 1 be bigger than angle 2. That's a contradiction! What did we do—what did we assume that got us into this mess?

What we did was we assumed the lines meet. So, if the lines meeting logically implies a contradiction, it's false that they meet. Therefore, the lines must be parallel, and we have proved it. If the alternate interior angles, angle 1 and angle 2, are equal—then the lines must be parallel. Good job, Euclid!

Notice what we needed logically in this proof and what we didn't need. We did not need the fifth postulate. Proposition 27—"If the alternate interior angles are equal, then the lines are parallel"—remains true no matter what might happen to Postulate 5. We did need Proposition 16 about the exterior angle, though.

You'll recall that I suggested that you notice in the proof of the theorem about the exterior angle being bigger than either opposite interior angle that we needed Postulate 2—that a line can be extended to any desired length. In that proof, we had a line, and we needed to double its length.

So the theorem I just proved—that if the alternate interior angles are equal, then the lines are parallel—that theorem also depends on Postulate 2. That fact is going to become important later on when we talk about non-Euclidean geometries. But for now, everything is working perfectly for Euclid.

Next, Euclid goes on to prove the 2 parts of Proposition 28—2 more facts about angles that will mean that the 2 lines are parallel. But now, we've got a way to prove that lines are parallel directly—all we have to do is show that the alternate interior angles are equal, and then we can invoke the theorem I just proved. So the proof of the next theorem, Proposition 28, is going to be a direct proof—no contradictions involved.

For the proof of Proposition 28, we're also going to need 2 theorems that we proved in Lecture Twenty-One. So let me restate them informally. These are about lines that meet. One of those theorems says: "If 2 lines intersect, the vertically opposite angles [like angle A and angle C in this diagram] are equal. The other theorem that I'm going to need says that: "When 2 lines come together [like this line and this line], the adjacent angles formed [like angle A and angle B in this diagram] add up to 2 right angles." Now we are fully equipped for the proof of Proposition 28.

Let's go and do it. For the first part: If what are called the corresponding angles—that's angle 2 and angle 3, and they're called corresponding angles because they're in positions that with respect to that third line, those positions correspond—if angle 2 is equal to angle 3, then the lines are parallel. We're given that angle 2 is equal to angle 3. Hey, we could prove the lines were parallel if only we knew that angle 1 was equal to angle 2.

Can we take the fact that angle 3 is equal to angle 2 and somehow make it tell us that angle 1 is equal to angle 2? Yes, we can.

Because, since angle 3 and angle 1 are vertically opposite angles, we know that angle 3 is equal to angle 1. If angle 3 is equal to angle 2, that's given, and angle 3 is also equal to angle 1. Hey, 2 things that are equal to a third thing are equal to each other, so they're both equal to angle 3. So angle 1 is equal to angle 2. Now we can invoke the theorem that I just proved, Proposition 27, and say now we've shown that the alternate interior angles are equal, and therefore the lines are parallel.

Here comes the second half of Proposition 28. For this part, we are given that angle 2 plus angle 4 add up to 2 right angles. Again, what we need in order to prove that the lines are parallel is to somehow turn what we're given into the conclusion that angle 1 is equal to angle 2—and we can do that. Remember that when 2 lines come together, the adjacent angles add up to 2 right angles. So, angle 1 plus angle 4 equals 2 right angles. That's from that theorem. But we were given that angle 2 plus angle 4 is equal to 2 right angles. So angle 1 plus angle 4 has got to equal angle 2 plus angle 4. Two things equal to a third thing are equal to each other—angle 1 plus angle 4 equals angle 2 plus angle 4.

Subtracting equals from equals will give you equals—that's one of Euclid's axioms. So if angle 1 plus angle 4 equals angle 2 plus angle 4, we subtract angle 4 from both sides—and that gives us angle 1 equals angle 2. We've got angle 1 equals angle 2, which, by Proposition 27, means that the lines are parallel—"If the alternate interior angles are equal, then the lines are parallel."

Notice, by the way, Euclid's proof of Proposition 28 needs Proposition 27; the proof of 27 needed the theorem about the exterior angle of a triangle, and the proof of that theorem needed Postulate 2. So the proof of Proposition 28 also rests on the truth of Postulate 2—that a line can be extended to any desired length.

As I keep repeating, because it's going to be so important when we get to non-Euclidean geometry, and as Plato said both in the *Republic* and in the *Meno*—geometry is hypothetical. Therefore, the proofs of Euclid's propositions hold only if the postulates are true. But given those postulates, Euclid's logic is just marvelous.

As Aristotle, Descartes, Spinoza, Newton, Thomas Jefferson, and lots of other people would agree, if you want truth and certainty, the thing to do is to construct a demonstrative science based on self-evident first principles—and logical deductions from them.

Let's turn from these proofs to the real world for a minute. Our immediate environment is just loaded with parallel lines. Look at a book end-on—the pages look parallel. Look at the sides of books ranged on bookshelves. Look at the bookshelves themselves. Look at the walls, and ceiling, and window frames in the room I'm in and whatever room you're in. Look at the lanes of traffic on the road, and the sides of buildings. Look at the sides of floorboards. Look at the sides of the hexagonal tiles on a bathroom floor. In all of these examples, we see the truth of Propositions 27 and 28, and we can use those theorems to construct these parallel lines.

Let me illustrate this last point with a story. One of my former students, incidentally now a published poet, he and I were walking together down the street in front of the college. He'd been having trouble with the idea that the relation between "if-then" statements in logic and the cause-effect relation are the same. He'd been having trouble with that.

Remember how Aristotle and Spinoza say that the relationship between premises and conclusions is like the relationship between cause and effect? Well, on this street that we were walking on, there were marked parking spaces, and all the cars parked at an angle—like this.

So I asked this student: "How do they make the lines in the parking spaces parallel? Do you think they draw some lines and then extend them for a mile or so to make sure they don't intersect?" "No, of course not," he said, "they've got to have some device that paints the same angle over and over again—like come along here and make this line this angle with the sidewalk, and then come along here and make this line the same angle with the sidewalk." "Hey," he said, "that's Proposition 28: 'If the corresponding angles are equal, then the lines are parallel'!" So he saw—we saw it together—how the premise: The corresponding angles are equal" is what causes the lines to be parallel: the conclusion. That was pretty exciting for both of us. So you can see the cause and effect operating as the guys paint the lines, like the: "If the corresponding angles are equal, then the lines are parallel."

Now we need to look at Euclid's Proposition 31, which proves that we can construct parallel lines. Because up to this point, we've proved some facts about parallel lines, but we haven't proved that there are any parallel lines.

Recall that to ensure that there'd be a straight line between any 2 points, Euclid had to assume it—that was Postulate 1. In order to ensure that a line could be extended as long as you want, Euclid had to assume that—that's Postulate 2.

Likewise, to ensure that we could construct a circle with any center and any radius, Euclid had to assume that—and that was Postulate 3. But—and here's an important point that I haven't emphasized very much so far—we want to assume as little as possible, even if you think your assumptions are clear, and distinct, and self-evident—otherwise, you might just as well assume the truth of every theorem in Euclid. So, except for the constructions I just quoted from Postulates I, II, and III that give us straight lines and circles, every time Euclid needs to construct something, he proves that it can be done. We saw that for the equilateral triangle, and that's going to be the case also with the construction of parallel lines. He'll prove that we can do it.

For his proof, he needs one more earlier theorem: that we can copy an angle and put it anywhere we want—more precisely: "With any point as vertex and any line as one side of the angle, you can copy the angle." I don't have time to prove that we can copy any given angle, so take it on faith—or look it up in Book 1 of the *Elements*; it's Proposition 23. But in any case, it's a proved result. We can copy an angle. Now we're ready for Proposition 31 that tells us we can construct parallels: "Given a straight line in the plane, and a point in the same plane not lying on the line—problem: to construct a line through that point parallel to the given line."

So we're given the point A and the straight line BC, and our mission is to construct a line through the point A parallel to the point BC. If the people who can make parking spaces do it, so can we. So we pick any point we like on the line BC. Let's call that point D—and then, by Postulate 1, we can connect that point D with A in a straight line.

Proposition 23 (that I mentioned before) said that we could copy any angle with a given line as one side, and a given point as the vertex. So let's copy the angle ADC, where the given line is AD and the vertex is A. I've just done that. I've copied this angle up here. So we have constructed angle EAD equal to angle ADC.

But look, if this angle, EAD, is equal to angle ADC, they're alternate interior angles, so Proposition 27 means that the lines EA and BDC are parallel. Hey, we've constructed a parallel line, there it is EA; we've constructed the parallel line EA parallel to BDC—and parallels now exist.

Great! Euclid has proved that parallel lines exist, and we can construct them. But this result raises at least 4 questions, which I will address in turn.

The first question, a logical one: Did we need the so-called "parallel postulate"—Euclid's Postulate 5—to show that parallel lines exist? The answer is "No." We did use Proposition 27, which presupposes Postulate 2—that a line can be extended to any length—but we did not use Postulate 5. We're going to need Postulate 5 in the next lecture, but not yet. Euclid has done quite a lot without Postulate 5.

Now the second question: "Hey, Professor Grabiner, I couldn't help noticing that you did the propositions in the order 27, 28, and 31? You skipped 29 and 30." By the way, that's one reason why I included the proposition numbers—so I could address this question.

I skipped Propositions 29 and 30 because I wanted to prove the ones that didn't need Postulate 5 first. But how come Euclid doesn't do it that way? How come Euclid proves 27 and 28, and then doesn't make the very next proposition the proof that parallel lines exist because they can be constructed (and doesn't need 29 and 30 to do that)? Euclid never tells us things like this, of course, but I think there is a plausible explanation.

Because, what did Euclid put in between Propositions 28 and 31 (between the facts about angles that make lines parallel and the actual construction of parallels he put something in between there)? What did he put in there? First, what is the proposition immediately before the construction of parallels? What is Proposition 30? Proposition 30 is the proposition that 2 lines parallel to a third line are parallel to each other.

What that means is that through any outside point here, like A, we can construct only one parallel to the line BC. That's because, if there were 2 parallels to the line BC through the same point, we'd have 2 lines parallel to a third line that meet, at A, so they would not be parallel to each other. So I think that, before proving that we can construct a parallel line, Euclid wanted to prove that the line we construct is going to be unique.

Now this fact—that through a given point there can be only one parallel to a given line—that fact is what is called "Playfair's axiom." You'll recall that I said—actually, Proclus said—that this fact is equivalent to Euclid's fifth postulate.

So how did Euclid prove Proposition 30, the uniqueness of parallels? He used—and he had to use—a result whose proof required Postulate 5. That result is Proposition 29. So that's why Proposition 29 also had to come before the construction of parallels.

We jumped from 28 to 31, but he has 29 and then 30. Proving Proposition 29—that states that: "If 2 lines are parallel, then the alternate interior angles are equal"—proving that one will be our first order of business in the next lecture.

Now for the third question, a deeper one. I said that the construction of a parallel in Proposition 31 shows that parallel lines exist. What does it mean to say that parallel lines exist? We're back to the fundamental questions raised at the very beginning of this course: Can you draw, on paper, 2 lines that are exactly parallel? No.

If not, where do parallel lines exist? If you don't believe in Plato's geometric heaven—that transcendent realm where the objects of mathematics supposedly live—are there alternative possibilities for where parallel lines exist?

They exist in the mind? Some modern neuroscientists say there's no such thing as "mind"—everything is in the brain. If they're right, do parallel lines exist in the brain? What does that even mean? Or do parallel lines exist in space? What is space? Is space something real?

All of these questions became especially urgent in the 18th century because, as we'll see in Lecture Twenty-Eight, Newton's physics seemed to require space to be real. You can try the dictionary. The dictionary defines space as: "something extended in which physical bodies exist."

That definition isn't much help in deciding whether space itself exists, or what kind of "something" space might be. In Lectures Twenty-Seven and Twenty-Eight, we'll see several important philosophers asking what space is and whether it exists—and, if so, where it exists. But right now, let's go back to Euclid, and let me ask my fourth question.

Assuming that parallel lines exist, at least in some sense of the word "exist," what are the major properties that we can prove that parallel lines have? I'll spend most of the next lecture working on this fourth question. Thank you.

# Lecture Twenty-Six
## Euclid—Parallels, Needing Postulate 5

**Scope:** In this lecture we come to the logical heart of the theory of parallels: Euclid's proof that, if 2 parallel lines are both cut by a third line, the alternate interior angles formed are equal. This result, part of Proposition 29, marks the only place Euclid uses his controversial fifth postulate explicitly. We then analyze the statement and proof of Proposition 30, that 2 lines parallel to a third line must be parallel to one another, and see how this depends on Proposition 29 and thus on Postulate 5. Finally, we discuss Proposition 32, that the sum of the angles of a triangle is 2 right angles and that the exterior angle of a triangle is equal to the sum of the 2 opposite interior angles. After proving this property, we return to the question "What is a rigorous geometric proof?"

## Outline

**I.** We now take up the 3 propositions in this section of the *Elements* whose proofs *do* require the fifth postulate, and we look at the logical relationship between these propositions.

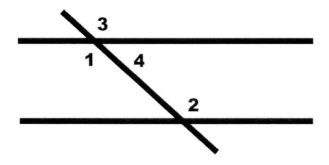

  **A.** Proposition 29 states that if 2 parallel lines are cut by a third line, the alternate interior angles (angles 1 and 2) are equal, the corresponding angles (angles 2 and 3) are equal, and the interior angles on the same side of the third line (angles 2 and 4) add up to 2 right angles.

  **B.** This is the fundamental theorem in the theory of parallel lines.

**C.** The proof of Proposition 29, as we will soon see, requires Euclid's fifth postulate (or some logically equivalent assumption).

**II.** We prove Proposition 29, following Euclid exactly.

    **A.** The first part, that angle 1 is equal to angle 2, is an indirect proof.

        **1.** We begin by assuming, without loss of generality, that angle 1 is greater than angle 2 (the argument where angle 2 is greater than angle 1 is exactly symmetric to this).

        **2.** We then show that this leads to the contradiction that the lines must meet—because of Postulate 5.

        **3.** If angle 1 is greater than angle 2, then angle 1 + angle 4 > angle 2 + angle 4, but since angle 1 + angle 4 = 2 right angles, angle 2 + angle 4 must be less than 2 right angles.

        **4.** If angle 2 + angle 4 < 2 right angles, then the lines must meet (by Postulate 5). But the lines are parallel, so we have reached a contradiction.

        **5.** Since the assumption that angles 1 and 2 were unequal led to a contradiction, they must be equal. So this equality—the most important part of the theorem—has now been proven.

        **6.** We repeat: This is the only place where Postulate 5 is used explicitly by Euclid, and the assumption he makes is exactly, no more and no less, what he needs to make this proof work.

    **B.** We now prove that angle 2 is equal to angle 3, using Proposition 15 about the equality of vertically opposite angles.

    **C.** Finally, we prove that angle 2 + angle 4 = 2 right angles, using Proposition 13.

**III.** We now prove Proposition 30.

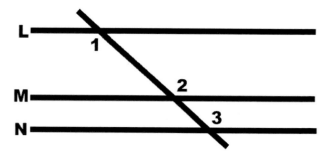

**A.** We have 3 lines—we call them L, M, and N—and are given that both L and N are parallel to M. We draw another line that cuts them all.

**B.** We use Proposition 29, whose truth depends on the truth of Postulate 5, to show that various angles are equal: alternate interior angles 1 and 2, and corresponding angles 2 and 3.

**C.** Using the equality of these angles, and Axiom 1 that 2 things equal to a third are equal to each other, we show that the crucial set of alternate interior angles, angles 1 and 3, are equal.

**D.** We invoke Proposition 27 to conclude that L and N are parallel.

**IV.** Finally, we prove the capstone theorem in this section, Proposition 32, that the sum of the angles of a triangle equals 2 right angles.

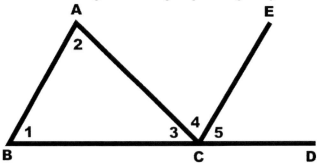

**A.** Before we do the formal proof, we show, by tearing off the corners of a paper triangle, that when we line up the angles they do look like 2 right angles.

    **1.** This procedure captures the intuition underlying Euclid's proof.

    **2.** Still, we imagine what Plato would say about such a "demonstration."

**B.** For Euclid's proof, we are given a triangle ABC; we extend the base BC to D, forming an exterior angle, and then, through the point C at that corner, we construct a line CE parallel to the opposite side AB.

    **1.** Note that these constructions are essential; the proof cannot proceed without them, a point to which we will return in Lecture Twenty-Eight.

    **2.** Extending the base requires the truth of Postulate 2.

> > > **3.** Constructing the parallel requires Proposition 31, which did not need Postulate 5, so at present we have not yet used that postulate.

> > **C.** Now we will need Proposition 29, which does depend on Postulate 5.

> > > **1.** Since CE is parallel to AB, angle 1 and angle 5 are corresponding angles, so they are equal.

> > > **2.** Angle 2 is equal to Angle 4, since they are alternate interior angles.

> > > **3.** Adding equals to equals, angle 1 + angle 2 = angle 4 + angle 5.

> > > **4.** Again adding equals to equals, angle 1 + angle 2 + angle 3 = angle 3 + angle 4 + angle 5.

> > > **5.** From Proposition 13, we know that angle 3 + angle 4 + angle 5 = 2 right angles. Therefore angle 1 + angle 2 + angle 3 is also equal to 2 right angles.

**V.** Some very important later results logically rest on these propositions:

> **A.** Proposition 47, the Pythagorean theorem: The square of the hypotenuse of a right triangle equals the sums of the squares of the other 2 sides.

> **B.** Figures with the same shape, also called similar figures, have proportional sides, be they triangles, parallelograms, or polygons.

**VI.** The logic of all these proofs is valid.

> **A.** As we know, therefore, the truth of the propositions so proven depends on the truth of the basic assumptions.

> **B.** Thus, the truth of a proposition could be challenged if a postulate turned out to be false.

> **C.** The assumption that historically was open to challenge was Postulate 5.

**VII.** But before the structure of Euclid's geometry was successfully challenged, the philosopher Immanuel Kant used it to try to solve one of the outstanding problems of philosophy.

**Essential Reading:**

Heath, *The Thirteen Books of Euclid's Elements*, vol. 1, Book 1, Propositions 29–30, 32.

**Suggested Reading:**

Heath, *The Thirteen Books of Euclid's Elements*, vol. 1, notes to these propositions, 312–315, 317–322.

Katz, *A History of Mathematics*.

**Questions to Consider:**

1.  Historically, Postulate 5 was challenged, with ancient geometers of the stature of Ptolemy and Proclus trying to prove it, and with Proclus calling it a "merely plausible and unreasoned hypothesis." If Euclid had, instead, assumed Playfair's axiom as Postulate 5, would it have been accepted as being at least as self-evident as Postulates 1–4?

2.  Is the theory of parallels given in Euclid's Book 1, Propositions 27–32, a good topic to use in teaching logic? Why or why not?

# Lecture Twenty-Six—Transcript
# Euclid—Parallels, Needing Postulate 5

Welcome back to lots of geometry. Let me promise you again: There will be quite a payoff. In this lecture, we're going to look together at 3 propositions about parallels whose proofs do require Euclid's Fifth Postulate. These are Proposition 30: "2 lines parallel to a third line are parallel to each other"; Proposition 32: "The sum of the angles of a triangle add up to 2 right angles"; and Proposition 29, which is the basis for both of the other ones—and thus supremely important to this theory. So we start with Proposition 29. Here it is.

Proposition 29 states: "If 2 parallel lines are cut by a third line, the alternate interior angles (that's angle 1 and angle 2) are equal; if 2 parallel lines are cut by a third line, the corresponding angles are equal (that's angle 3 and angle 2); and if 2 lines are parallel, the interior angles on the same side (that's angle 2 and angle 4) add up to 2 right angles."

As you can see, this proposition is the converse of 2 of the theorems I proved in the last lecture. As you'll recall, the converse of a statement does not logically follow from the statement itself. The converse of a true statement can be true or it can be false. Let's review this.

For instance, for the statement: "If you're in New York state, then you're in the Empire state"—the converse is true: "If you're in the Empire state, then you're in New York state." Or the converse could be false, as: "If you're in New York City, then you're in New York state"—that's true, but the converse: "If you're in New York state, then you're in New York City"—that's false. But even if a converse is true, it requires a proof of its own.

That's the case here. Proposition 29 is true in Euclid's system, but it needs its own proof, and—and this is the key point—that proof requires something that the proofs of the earlier theorems about parallel lines did not. Euclid's choice for the additional thing the proof of Proposition 29 requires is his Postulate 5. Because Proposition 29 is the crucial underpinning for the whole subject of parallels and for similar geometric figures—that is, the theory of same-shape geometric figures—Postulate 5 is equally crucial.

Since the proof of Proposition 29 is the only proof where Euclid explicitly uses Postulate 5, I've claimed that Euclid knew that this postulate was a problem and used it only when he absolutely had to. But Euclid acted as though there were no alternative to assuming this postulate. He evidently thought that it couldn't be proved from his other postulates—and, as it turned out, he was right.

Let's see how Euclid used Postulate 5 to prove his Proposition 29. In a way, the overall strategy of this proof will remind us of the proofs of Proposition 27 and Proposition 28 from last time. First, Euclid treats the case of the alternate interior angles—for that part, he gives an indirect proof. Then, he'll use the propositions about what happens when 2 lines intersect to get the results that are about the other relevant angles.

So here we go. What we want to prove is that if the 2 lines are parallel, then angle 1 is equal to angle 2. It's a proof by contradiction, so we're going to begin by assuming that angle 1 is not equal to angle 2. Without any loss of generality, we can assume that angle 2 is less than angle 1. The proof for the other case, when angle 1 is less than angle 2, is exactly the same; all we have to do is turn the diagram upside down. So I'm going to do this—just this half.

So we can assume that angle 2 is less than angle 1. This means that angle 2 plus angle 4 is less than angle 1 plus angle 4. But we know what angle 1 plus angle 4 is—that's 2 right angles—180 degrees, if you like—by the theorem that says that when 2 lines come together, the adjacent angles here (angle 1 and angle 4) add up to 2 right angles. We proved that back in Lecture Twenty-One. What that means then is that angle 2 plus angle 4 are less than 2 right angles—angle 2 plus angle 4 add up to less than 2 right angles.

Hey! We know what happens in that case! Postulate 5 tells us! Postulate 5 says that: "If 2 lines are cut by a third line such that the interior angles on the same side of the third line add up to less than 2 right angles, then the lines meet." But wait! The lines can't meet. This proposition is about what happens to the angles when 2 lines are parallel. Two lines can't both meet and be parallel—the definition of parallel lines is that they don't meet. So we have reached a contradiction.

Ladies and gentlemen of the jury: We are asked to believe that if 2 parallel lines are cut by a third line, the alternate interior angles are not equal. But if they aren't equal, it follows that these parallel lines meet! That can't happen. So angle 2 can't be less than angle 1. By exactly the same argument, we can show that angle 1 can't be less than angle 2. So, angle 1 has to equal angle 2. I rest my case. The logic is perfect—but we needed Postulate 5 to make this proof work.

Now we've done the hard part, so let's prove the rest of Proposition 29. First, we need to prove that if 2 lines are parallel, then the corresponding angles are equal—that is, angle 2 is equal to angle 3. That's pretty easy. We just proved that if 2 lines are parallel, then angle 1 is equal to angle 2; angle 1 is equal to angle 3 also—they're vertically opposite angles. So, if 2 lines are parallel, angle 2 is equal to angle 1, and angle 3 is equal to angle 1, so angle 2 is equal to angle 3 because 2 things equal to a third thing are equal to each other. So, we now have that if 2 lines are parallel, the corresponding angles—angle 2 and angle 3—are equal. To get angle 2 equals angle 3, we needed a result—if the 2 lines are parallel, then angle 1 equals angle 2—that depended on Postulate 5. So this new result—if 2 lines are parallel, then the corresponding angles are equal—also depends on Postulate 5.

Finally, we can prove that angle 2 plus angle 4 has to be equal to 2 right angles. If 2 lines are parallel, then angle 2 plus angle 4 is 2 right angles. Since we already have, if 2 lines are parallel, angle 1 is equal to angle 2, we can add angle 4 to both sides of that equation, because if equals are added to equals, the sums are equal—that's an axiom. So angle 1 + angle 4 = angle 2 + angle 4.

But angle 1 plus angle 4 is 2 right angles—again by that theorem that: "When 2 lines come together, the angles add up to 2 right angles." So, if my 2 original lines are parallel, then angle 2 plus angle 4 also equals 2 right angles, and that's what is required. This result also needed angle 1 is equal to angle 2—so this result also logically depends on Postulate 5.

Euclid must have known that the proposition I've just proved can also be proved using what we know call "Playfair's axiom"—the uniqueness of parallels—instead of Postulate 5. We'll never know for sure why Euclid chose his own Fifth Postulate instead of assuming the uniqueness of parallels—but it's kind of interesting to speculate about this. One reasonable guess that people have made is that if you prove that if 2 lines are parallel, then the alternate interior angles are equal; by using the uniqueness of parallels, the proof requires an additional construction. But any additional complication—like making another construction—makes you worry that the proof might be making some implicit assumption we don't even realize we're making. Instead, Euclid's proof is about as clear and clean as you could imagine. Postulate 5 is exactly what he needs to make his proof so clear and clean.

There's a downside to that, though. The downside is that Euclid's Postulate 5 is less self-evident than Playfair's axiom. Playfair's axiom has a kind of self-evidence about it—given a line and an outside point there is only one parallel. So here's something else to wonder about: Suppose Euclid had chosen the uniqueness of parallels as his Fifth Postulate—that is, Euclid had explicitly assumed instead of his Postulate 5 that there can be only one parallel to a given line through a given outside point. That might have seemed much more self-evident. If Euclid had done this, maybe people wouldn't have thought: "Hey, this is something that needs proof," so they wouldn't have spent 2000 years trying to prove it, and they wouldn't have gotten frustrated by their lack of success—because it was that lack of success, that frustration, that ultimately led to the realization that there could be a different, a non-Euclidean, geometry in which Postulate 5 did not hold.

So maybe if Euclid's chosen parallel postulate had been more self-evident, non-Euclidean geometry might never have been invented at all. As I said, it's interesting to wonder about—but, of course, there is no way to know. But be that as it may, we have now finished proving Proposition 29, and I will use it again and again as I prove the other 2 propositions in this section of the *Elements*. First, Proposition 30—it is kind of nice: "Two lines parallel to a third line are parallel to each other."

This essentially proves Playfair's axiom, the uniqueness of parallels. Here's how Euclid's proof goes: Suppose we have 3 lines that I'm going to call L, M, and N. Suppose we are given that L is parallel to M and that N is parallel to M. So, we've got these 2 lines that are parallel to a third line. What we want to do is prove they are parallel to each other—to prove that L is parallel to N. How are we going to do that? First, we draw another line that cuts all 3 of them, and we label some of the angles, as I show you here. Since L is parallel to M, angle 1 is equal to angle 2, because they're alternate interior angles. Proposition 29—right? "If 2 lines are parallel, then the alternate interior angles are equal."

Likewise, since N is parallel to M, angle 3 is equal to angle 2 since they're corresponding angles—that's Proposition 29 again. But since both angle 1 and angle 3 are equal to the same angle 2, they are equal to each other (angle 1 is equal to angle 3). But look back at the diagram. With respect to the lines L and N, angle 1 and angle 3 are alternate interior angles. Our friend, Proposition 27, told us that: "If the alternate interior angles are equal, then the lines are parallel." So L must be parallel to N, and that is what Euclid set out to prove.

The uniqueness of parallels is, as I've said repeatedly, equivalent to Euclid's Postulate 5. So it's no surprise that the proof of the uniqueness of parallels depends on Postulate 5—and we see that it does, because this proof needed Proposition 29.

We now turn to Proposition 32—that: "The sum of the angles of a triangle is 2 right angles." Descartes and Spinoza used this—"The sum of the angles of a triangle is 2 right angles"—as a model theorem when they talked about geometrical truth. Why is this such a good example? Why did these guys use it?

Because, although its truth is not obvious from just looking at a diagram of a triangle, once you've seen the proof, there is no longer any room for doubt. I mean that's supposed to be the whole point of proof-based geometry—that we get new, often unexpected results, and then we prove that they are true.

But before I set out to prove this, let me ask this: What makes us think that it's true? I mean before you set out to prove something, you need to know—or have some idea, anyway—that it's true. We can convince ourselves that this might be true by a physical experiment.

It's kind of hard to visualize the sum of 3 angles when the angles aren't all together—all around the same vertex, but they're in 3 different places. We were fine in Proposition 13, when there were intersecting lines, when the angles we were talking about were right next to each other, but it's kind of hard to visualize bringing angles together from different parts of a diagram.

So to make sense out of this, let me do an experiment that physically assembles the 3 angles of a triangle, and let's see if they do add up to 2 right angles.

Here is a paper triangle. What I'm going to do is I'm going to tear off 2 of the angles of this triangle, this one and this one, and I'm going to re-assemble them—together with the third angle that I've left in place—to see if they line up along a straight line. If they do lie along a straight line, then we know from Proposition 13 that they add up to 2 right angles. I'm going to show you. Let's tear off angle 1 and put it over here. Let's tear off angle 2 and put it over here. I've got to get this in the right place. Look at that folks. They lie along a straight line—so they do, in fact, add up to 2 right angles.

In fact, this little demonstration in a way captures the intuition that underlies Euclid's proof of the theorem that the sum of the angles of a triangle is 180 degrees or 2 right angles, as you'll see when I do the proof. But Plato is turning over in his grave. "Come on!" he says, "a triangle isn't made of paper. You can't really, no matter how hard you try, you can't line the angles up perfectly."

You can't tell if they add up to 2 right angles or 1.99 right angles. All you've got there is a paper image of the real triangle that the theorem is about. Plato would go on saying: "Well, maybe this little demonstration has helped to prepare you. But now, I want you to come out of the cave, and let's look at this question with the eye of the intellect."

We can re-assemble this triangle. Let me just run this by one more time. We picked up this angle, and we put it over here. We picked up this angle, and we put it over there. There they are, angle 1 plus angle 2 plus angle 3—2 right angles. All right? Okay, Plato, we will retire this. Now we're going to do it right. So, theorem: "The sum of the angles of a triangle is 2 right angles—or if you like, 180 degrees." So we are given a triangle, ABC. What we want to prove is that angle A plus angle B plus angle C add up to 2 right angles.

The first step in Euclid's proof is a construction.

We extend the base BC to D. That forms the exterior angle ACD.

Through the point C, which is at the corner of that exterior angle, we construct a new line—the line CE, which is parallel to the other side of the triangle. Let me make this explicit: CE is parallel to AB. You know we can do that, right? Remember the theorem that said we knew how to construct parallel lines?

These constructions are essential. The proof cannot proceed without these constructions. That's an important point, and it's going to become even more important philosophically later on. Extending the base BC, that's okay because of Postulate 2—and constructing the parallel line CE is legitimate because Euclid proved we can construct parallels in Proposition 31. Since Proposition 31 didn't require Postulate 5, we haven't used Postulate 5 so far—but that's about to change.

Now I'm going to argue that, because CE is parallel to AB, there are a lot of angles that are equal. That will require using Proposition 29—twice. Proposition 29 depends on Postulate 5, so our proof will also logically depend on Postulate 5. All right, that's enough introduction. Here goes Euclid's actual proof.

Because the line EC is parallel to the line AB, angle 1 and angle 5 are corresponding angles—so they are equal to each other; angle 1 is equal to angle 5—and (this is a little harder to see immediately—it's going to help to tilt your head a little bit) because EC is parallel to AB, angle 2 is equal to angle 4.

Angle 2 and angle 4 are alternate interior angles—so since those lines are parallel, angle 2 and angle 4, being alternate interior angles, they are equal. We can add equals to equals, that gives us equal results—so by taking those 2 pairs of equal angles, we add them up: angle 1 plus angle 2 is equal to angle 4 plus angle 5.

Let's stop and look at that result geometrically for a minute: angle 4 plus angle 5, added up together, are the exterior angle ACD. So angle 1 plus angle 2 is equal to angle ACD.

You remember Proposition 16? Proposition 16 said that the exterior angle was greater than either opposite interior angle—so that proposition said that "Angle ACD is greater than angle 1, and angle ACD is also greater than angle 2." What Euclid has just shown, though, is a much stronger result. Instead of an inequality, we have an exact equality—but, of course, we're assuming more. We're using Postulate 5. So it's not surprising that with more assumptions we get a stronger result—and Euclid isn't done.

Since angle 1 plus angle 2 is equal to angle 4 plus angle 5, we can add the same thing to each side of this equation, and it will still be true. So, let's add angle 3 to both sides of this equation. So, angle 1 plus angle 2 plus angle 3 is equal to angle 4 plus angle 5 plus angle 3. But—and here is something I showed you when I used that paper triangle—angle 3 plus angle 4 plus angle 5 all lie along the same straight line.

That means that their sum—angle 3 plus angle 4 plus angle 5—is equal to 2 right angles by our often-used Proposition 13. So, since 2 things equal to a third thing are equal to each other, angle 1 plus angle 2 plus angle 3 also has to equal 2 right angles. Notice that if we had torn off the top of the triangle (let's have another quick look), if we had torn off the top of the triangle and put angle 2 down on top of my angle 4 and torn off the left side and slid angle 1 over and put it on top of angle 5, we would have reproduced our little experiment on Euclid's diagram and showed that the physical angles look as though they add up to 2 right angles.

So that physical experiment really does give us the intuition that underlies Euclid's proof. But again, intuition is all very well, but the goal of Greek geometry is to achieve certainty based on logical deductions from explicit basic assumptions. That goal is not met by ripping up physical triangles; it's met only by Euclid's demonstrations—and what a demonstration that is. The logic of this is just perfect.

Not everything in Euclid is so perfect. You'll recall from Lecture Sixteen that in his very first theorem, he unknowingly used an assumption that he didn't state. But there are no such gaps in the proofs about parallels in this and the previous lecture.

Euclid recognized exactly what had to be assumed, and he explicitly assumed those things. So, later mathematicians looking at this could see exactly where Postulate 5 was essential and where Postulate 5 was not essential.

In Euclid's geometry, the propositions that I've proved today are the basis for many, many important results later on. Let me state just 2 of them. The proofs of both of these results require earlier results based on earlier results, and so on—ultimately based on Proposition 29, so these both depend on Postulate 5.

First is the Pythagorean Theorem: "In a right triangle, the square on the hypotenuse is equal to the sum of the squares on the other 2 sides." Euclid's proof of this theorem, by the way, is just beautiful. Unfortunately, it would probably take a whole lecture for me to work through it all, so I'm not going to do it. But if you're interested, I highly recommend working through it. It's Book 1, Proposition 47. It's a beautiful proof, particularly because it uses things that don't seem like they are going to be related, but that come into the proof in exactly the right time and the right place. It's really, really nice.

The second result I want to highlight here is the result that figures with the same shape—also called "similar figures"—have proportional sides. Let me say that again: Figures with the same shape—also called similar figures—have proportional sides. That's true whether the figures are triangles, or parallelograms, or polygons. For instance, in Euclid's Book 6, Proposition 2, we have the result for triangles: Similar triangles have proportional sides. That is, suppose we're given the triangle here, ABC.

If we draw the line DE parallel to the base BC, then this little triangle has got the same shape as the big triangle, and we can show that—okay? If you draw this DE parallel to base BC, the corresponding angles at D and at B are equal—and at E and at C are equal. They share angle A, so they are all the same shape. So the little triangle ADE and the big triangle ABC have the same shape; they are similar triangles.

Then Euclid proves (I'm not going to prove it, but I definitely want to state it) that AB is to AD as AC is to AE. Whatever the proportion is between AB and AD is the same proportion between AC and AE. That result, as we will see in Lecture Twenty-Nine, is the central fact in the geometric theory of perspective. So this result is very important in the history of art.

Let me sum up now the nature of Euclid's achievement about parallels. The logic of all of these proofs is valid. So the truth of the propositions Euclid has proved depends on the truth of the basic assumptions. That means that the truth of one of these theorems could be challenged if one of the postulates turned out to be false. There was one assumption that, historically, had been challenged—and that assumption was Postulate 5.

Euclid has laid out the logical structure of the theory of parallels honestly and clearly. He needed to assume something like Postulate 5, and he did assume it. So here are the major questions that remain. Given the truth of the rest of the postulates and the truth of the axioms, is it possible to use them to prove that Postulate 5 is true? Or, might it be possible to use them—the other postulates and axioms—to prove that Postulate 5 is false? These aren't the only choices, either. Could Postulate 5 be independent of all the others, so that there could be one geometry that assumes Postulate 5 to be true and all the rest of the postulates to be true, and another geometry that assumes all the rest of the postulates to be true, but assumes that Postulate 5 is false?

When we come to talk about non-Euclidean geometry in Lectures Thirty and Thirty-One, we'll see how those questions finally got answered. But that's getting a little ahead of our story. It was a long time before Euclid was challenged—let alone successfully challenged. For over 2000 years, as I've said so many times, Euclid's geometry stood as a model of correct reasoning and true conclusions.

But it could be even more than a model of good reasoning? The great philosopher Immanuel Kant used the ideas behind Euclid's geometry to try to solve one of the most important problems in philosophy. I'll just state the problem—I'll explain what it is in the next lecture, but the problem is: Is metaphysics possible?

In the next lecture, we'll see how Kant did this. Now we are very well prepared to understand Kant's argument. I think it's fair to say that many readers of Kant do not completely "get" the mathematical dimension of Kant's argument—but we will.

So, in the next lecture, we'll examine this historically significant episode in the history of philosophy. We will find that understanding Euclid's theory of parallels in general and the theorem that the sum of the angles of a triangle is 2 right angles (the proof of that theorem in particular), that this knowledge helps us understand Kant's profound arguments about space, time, causality, and metaphysics. Thank you.

# Lecture Twenty-Seven
## Kant, Causality, and Metaphysics

**Scope:** In the next 2 lectures, we look at Immanuel Kant, the philosopher most profoundly influenced by Euclid's theory of space. Kant is often considered the defining philosophical figure of the Enlightenment. In this lecture, we will look at the way Kant addressed the question "Is metaphysics possible?" and at what his answers owe to Euclid's geometry. Kant argued that the traditional answers to metaphysical questions like "How can we tell whether every effect has a cause?" are all wrong. Kant provides a way of classifying types of statements, with statements like "every effect has a cause" falling into one of his classifications. Kant asks whether there are any examples of statements that clearly are of this type. If there are not, maybe the whole program of creating metaphysics is hopeless. But if there are, we know that *some* statements with the properties we want for the truths of metaphysics actually do exist. Will geometry provide examples of such statements? Stay tuned.

## Outline

I. Our theme throughout these lectures has been the relationship between mathematics and philosophy, and when people think of philosophy, they often are thinking about the branch called metaphysics.

   **A.** Metaphysics deals with those questions about reality that transcend any particular science.

   **B.** Here are some examples of metaphysical questions:
   1. What exists, if anything, beyond the world of sense perceptions?
   2. Are there laws of nature?
   3. If there are laws of nature, is it necessary that they be exactly as they are?
   4. Must every effect have a cause?
   5. What are space and time?
   6. Do space and time exist independently of our ideas of them?

   **C.** For the purpose of the present lecture, the most important property of metaphysical questions is that they are not investigated by the methods of empirical science.

**II.** Near the end of the 18<sup>th</sup> century, the perceived triumph of Isaac Newton's physics still left a number of these questions unanswered.

    **A.** Newton and his followers held that space was real. The reality of space was vital to Newton's concept of force.

    **B.** Gottfried Wilhelm Leibniz, however, denied that space was real, saying instead that all that existed was the relations between the physical bodies.

    **C.** The skeptical philosopher David Hume went so far as to deny that there was causality in the world.

**III.** Immanuel Kant came down firmly on the side of causality, and he therefore believed that all events follow determined laws.

    **A.** Kant's argument for this makes it clear that this is not a question of science, but a question of metaphysics.

    **B.** The urgency of the questions about causality motivated Kant to ask whether metaphysics was possible.

    **C.** To answer this question, Kant strove to find out what kind of statements made up metaphysics, and whether we had any reason to believe that there were any statements of this kind.

    **D.** Kant wrote his *Prolegomena* [preparatory exercises or observations] *to Any Future Metaphysics* to investigate whether propositions that transcend any particular experience could exist, and if so, how one could come to know their truth.

**IV.** In his writings, Kant focuses on the kind of proposition he calls a judgment, and he classifies judgments according to how one comes to decide their truth.

    **A.** A judgment is a statement of the form "A is a B," or as he puts it, subsuming a particular under a universal. Examples include "The grass is green" or "A bachelor is unmarried."

    **B.** We can come to know the truth of some judgments just by analyzing the terms in them.

    **C.** A judgment whose truth is determined just by analyzing the terms in it is called an analytic judgment.

    **D.** A judgment whose truth cannot be determined just by analyzing the terms, so that to decide its truth we must appeal to something outside the terms of the judgment, is called a synthetic judgment.

**E.** To determine the truth of a proposition that is a synthetic judgment, one can appeal either to sense experience or to something other than sense experience.

    **1.** Judgments whose truth depends on sense experience, or empirically based judgments, are said to be a posteriori judgments.

    **2.** Judgments whose truth can be determined independently of sense experience are called a priori judgments.

**F.** We have 2 sets of 2 categories, so the multiplication principle says that there are 4 possible types of judgments:

    **1.** Analytic a priori.

    **2.** Analytic a posteriori.

    **3.** Synthetic a posteriori.

    **4.** Synthetic a priori.

**G.** We comment on each possibility in turn:

    **1.** Analytic a priori judgments obviously exist; any judgment that is true by definition qualifies, ranging from "a bachelor is unmarried" to "at least half the scores are at or above the median."

    **2.** Analytic a posteriori judgments do not exist, since they involve a contradiction in terms. If all we need to do is analyze the terms, we do not appeal to sense experience to find out whether the judgment is true.

    **3.** Synthetic a posteriori judgments also exist. Any nontrivial judgment about the world of sense experience requires an appeal to sense experience, whether it is "The grass is green" or "Mars goes around the Sun in an orbit shaped like an ellipse."

    **4.** Now for the hard one: Do synthetic a priori judgments exist?

**H.** Turning the question around, we ask, "In which of these categories are the judgments of metaphysics?"

    **1.** Kant makes clear that they cannot be a posteriori.

    **2.** For instance, we cannot determine whether nature always follows laws by appealing to experience and observation.

    **3.** Nor are the judgments of metaphysics analytic, because we do not define nature as following laws, nor do we define facts (effects) as things that must have causes. Whether they do or not is exactly the kind of question metaphysics is supposed to answer.

    **4.** Therefore, the judgments of metaphysics must be synthetic a priori.

**V.** So, key questions and goals in Kant's *Prolegomena* are these:

    **A.** Do synthetic a priori judgments exist?

    **B.** If so, are there any synthetic a priori judgments whose truth is unanimously recognized?

    **C.** If so, how do we come to know that they are true, or as Kant puts it, "How is synthetic a priori knowledge possible?"

    **D.** If there are such judgments, and we can determine their truth by reason, metaphysics is possible.

    **E.** And as we will see in the next lecture, such judgments do exist: in geometry.

**Essential Reading:**

Russell, *A History of Western Philosophy*, 701–718.

**Suggested Reading:**

Kant, *Prolegomena to Any Future Metaphysics*.

Koerner, *Kant*.

**Questions to Consider:**

**1.** Why should anybody care about these metaphysical questions?

**2.** If Niels Bohr, James Clerk Maxwell, and others who believe that reality is essentially statistical in nature are correct, this means that the statement "Every effect has a cause" is not true. What, then, is the status of the judgment "Reality is essentially statistical"? Think about this both from the perspective of Kant's classification of judgments and from other points of view.

# Lecture Twenty-Seven—Transcript
## Kant, Causality, and Metaphysics

Hello again. At the start of this course, I promised that I would talk about how mathematics has influenced philosophy. The examples so far—Plato, Aristotle, Descartes, Spinoza—have not been obscure figures. In the next 2 lectures, I'll focus on the giant among philosophers of the Enlightenment: Immanuel Kant.

Kant is a towering figure in many areas of philosophy—from ethics to aesthetics. But often, when people think about philosophy, the questions that interest them most are from the branch called "metaphysics." In my opinion, metaphysics is the area of Kant's greatest achievements.

So what is metaphysics anyway? Metaphysics deals with those questions about reality that transcend any particular science. Here are some examples of metaphysical questions: What exists, if anything, beyond the world of sense perception? Does nature follow laws? If there are laws of nature, is it necessary that those laws be exactly as they are—or could they be different? Must every effect have a cause? Or do some things just "happen"? What are space and time? Do space and time actually exist, independently of our ideas about them? What all of these questions have in common is that the way to investigate them is not by using the methods of empirical science.

We've already seen a number of philosophers and scientists talking about these questions: Laplace and Maxwell, Einstein and Niels Bohr, Plato and Aristotle, Descartes and Spinoza, Newton, Leibniz, Hume. But, as I hope I can convince you, Kant is special.

I want to start this introduction to Kant's metaphysics by explaining why the questions about what space is, and whether space is real, were so important in the 18th century. Then, I'll say just a few words about debates about causality in the 18th century. After that, I'll move back to metaphysics in general and address the fundamental question that Kant posed about metaphysics, which is: "Can we ever have knowledge about questions like these?"—questions that are here on this slide. Or as Kant put it: "Is metaphysics possible?" In this lecture, I'll explain the ideas that Kant developed to help address the question: Is metaphysics possible? In the lecture that follows, Lecture Twenty-Eight, we'll see how Kant tried to answer the question.

First, though, for space—I want to start with the question about whether space is real, and I'll start with Newton. Newton and his scientific followers did say that space was real. Why? Well, suppose you're sitting in a car that is really accelerating toward another car. You are pushed back into your seat by the force that that acceleration produces. It may look, as you look through the windshield, as though the other car is accelerating towards you—but you know that it's you who is accelerating because of the force on you.

If you're sitting in a car that isn't moving, and another car is accelerating toward you, what you see—the front of the other car appearing to accelerate toward you—what you see looks exactly the same as what you saw when you were accelerating towards it. But you don't feel a force pushing you back into your seat. So you can tell that the other car is the one that's really accelerating, not you—really accelerating, but really accelerating with respect to what?

In both cases, each car appears to be accelerating with respect to the other car. But Newton would say: "Look, in the first case, your car is really accelerating with respect to space—and you know that because the acceleration—the real acceleration with respect to space—produces a force, a force that pushes you back in your seat."

In the second case, it's the other car that's accelerating—while your car, with respect to space, is at rest. Again, you can tell that your car is not accelerating with respect to space, because the apparent acceleration of your car produces no force on you.

Of course, Newton's great achievement was the law of gravity. Newton wanted to talk about the force of gravity, and he wanted gravity to be a real force—and, for Newton, real forces produce real accelerations. Let me say that again. Real forces produce real accelerations. So, Newton needed absolute space as a reference frame so that he could establish that there was a difference between real and apparent accelerations: Real accelerations need a force; apparent accelerations don't.

Newton's Second Law of Motion gives the precise relationship between force and acceleration. In modern notation: $F = ma$—force equals mass times acceleration. That means that real acceleration produces real forces. To really convince you that you need this idea of acceleration with respect to absolute space, let me give another example. There's an amusement park ride where they put you inside of a room shaped like a cylinder. You have your back to the wall of the cylinder, and then they start the cylinder spinning around really fast. You get pushed. You feel plastered against the wall of that cylinder. Then, they drop the floor out from under you—no floor! But you don't fall because the force from the spinning keeps you pushed back against the wall. Finally, first they put the floor back, and then they stop the ride. They'd better do it in that order!

Now imagine that this spinning cylinder in this ride is inside of a big tent. Before, the cylinder was spinning with respect to the tent. Now let's do something different. You're still standing against the wall inside the cylinder—but this time, let's leave the cylinder at rest and rotate the big tent around it. Okay, now the tent is spinning fast around the cylinder. Now would you let them drop the floor out? I don't think you would.

What you see is the same—you see the relative spinning of the cylinder and the spinning of the tent. Okay? But—and here's what Newton would say: "When the cylinder isn't really spinning with respect to absolute space, there's no force on you—so you don't want them to drop the floor."

Newton needs this idea of real space so he could establish that the forces involved with true accelerations, as opposed to relative accelerations, are real—because he wants the force of gravity to be real; otherwise, gravity can't explain the universe. Newton needs real space for real forces. What he says is: "Absolute space, in its own nature, without regard to anything external, remains always similar and immovable.

By contrast, Leibniz did not believe in absolute space. Leibniz said that spatial relations were just the relations between bodies. We don't need an outside framework to see where you are with respect to me. In fact, Leibniz used a metaphysical principle—his principle of sufficient reason—to argue that there was no such thing as absolute space. If there were absolute space, Leibniz said there would have to be a reason to explain why these 2 objects would be related in one way if East was over here and West over there—and related in a different way if East and West were reversed.

Surely, Leibniz said, the relation between 2 objects is just one thing. Leibniz would explain the spinning cylinder experiment by saying that what Newton was calling "acceleration with respect to space" is, in fact, acceleration measured with respect to all the physical bodies in the universe—not measured with respect to some fictional abstract space. That's what Leibniz would say. So this metaphysical issue—"Is space real?"—has important implications for physics. It's not just an abstract philosophical argument.

Another metaphysical issue in the 18<sup>th</sup> century is about causality and natural laws—and here we have Aristotle, and Spinoza, and Laplace saying: "Definitely, yes, every effect has a cause. Our science tells us what causes what, and all natural phenomena follow definite laws."

But as we saw back in Lecture Ten, we have Maxwell insisting that there's always a little element of chance and freedom. In the 18<sup>th</sup> century, we saw David Hume denying that there is any causality at all. To use a modern example: Flip a switch; the light goes on; we can't be certain that the first caused the second, any more than the rooster has the right to assume that his crowing causes the sunrise. Borrowing the language of Stephen Jay Gould, I would say that Hume argues that talking about causality is just a way of reifying an association that we have habitually encountered.

Where is Kant on this issue? Kant comes down firmly on the side of causality—and, therefore, believes that all events follow determined laws. I'll tell you part of his argument for this—not because I want to convince you that he's right, but because his argument on behalf of the principle "nature always follows laws" will make absolutely clear that "nature always follows laws" is not a question of science, but a question of metaphysics.

That is to say, we can't answer the question: "Does nature always follow laws?" by the methods of empirical science. Suppose I want to do that— suppose I want to claim: "Everything follows laws." You rightly ask: "What evidence do you have for this?" I say: "Well, the evidence is all the scientific laws we have." "Well," you object correctly, "there are lots of phenomena that we don't have any laws about."

But I've got to reply. I say: "Sure, but that's just because the laws haven't been discovered yet." You say: "Oh, yeah." Neither of us has enough evidence to settle this question. In fact, that little exchange makes really clear that no empirical investigation can settle this question. So the question: "Does nature follow laws?" belongs to metaphysics.

At this point, some hard-headed people—like David Hume—might say: All this metaphysics is just moonshine. We're never going to be able to answer questions like: "Do all natural phenomena follow laws?" or "Does every effect have a cause?" or "What are space and time, and are they real?" We're never going to be able to answer those questions. All we have is what we get from our senses: sight, hearing, touch, and so on. We can reason about our sense perception—we can reason about our ideas—but these sense perceptions are the only reality. But since Newton's physics seem to need absolute space, and since anybody who's doing science kind of wants to believe that there really are natural laws out there waiting to be discovered—these metaphysical questions seem to be very important.

So, what I've just said re-creates the historical situation in which Kant found himself near the end of the 18th century. But Kant didn't jump into this situation and say he had definitive answers. His approach was more modest. Instead of answering questions like "Does every effect have a cause?" he starts in a more modest way: "Let's not ask the question; let's instead ask whether the subject that would answer such questions—metaphysics—whether that subject could exist at all."

To answer that question—"Is metaphysics possible?"—Kant had to come up with a very precise way of explaining exactly what kinds of statements the statements of metaphysics were. Kant is famous for his precise, but not obvious, use of words. This makes Kant kind of hard some people think. But once you've mastered Kant's terminology, I think you'll find him clear. So don't be put off by his technically precise—but unfamiliar—terms.

The rest of this lecture may at first seem abstract and difficult—partly because of Kant's terminology. But if you persevere, I think you'll be rewarded as you master Kant's deep insight into what he thought was the nature of geometry, space, and time.

Kant wrote a book called *Prolegomena to Any Future Metaphysics*. *Prolegomena* is a Greek word in plural form, and it just means "preparatory exercises or observations." People who talk about this little book go around calling it the *Prolegomena*, so it's worth recognizing the term. Now to the action. Kant's *Prolegomena* sets out to answer the question: "Is metaphysics possible?" To do this, Kant decided that he had to investigate these 2 questions: What kind of judgments are the judgments of metaphysics? Do we have any reason to believe that any judgments of this kind exist?

All right, so, here is our first technical term: "judgment." A judgment, for Kant, is a statement of the form "A is a B"—or, as he puts it: "subsuming a particular under a universe." Here are some examples: "The grass is green"; "A bachelor is unmarried"; "Socrates is rational." All right, now we know what a judgment is. Now let's ask: What kind of judgments are the judgments of metaphysics?

What does he mean: What kind of judgments? Kant classifies judgments according to how we come to decide whether the judgments are true or false. So, then, how do we decide whether a judgment is true or false? What do we appeal to? We can figure out the truth or falsity of some judgments just by analyzing the terms in them. For instance, consider: "This golden ring is golden." We know that that's true, and we don't need a jeweler to help us. We know that "This golden ring is golden" is true just by analyzing the term "golden ring."

Another example is "A bachelor is unmarried." We analyze the term "bachelor," and we find "unmarried man." So we can decide that this is a true judgment without going down to the marriage license bureau and looking the guy up. Likewise, if somebody says: "This bachelor is married"—we can tell that that's false just by analyzing the terms. Kant has a name for this. He calls judgments like this "analytic judgments."

So, here is the first really important technical term: "A judgment whose truth is determined just by analyzing the terms in it, with no reference needed to anything else, is called an analytic judgment."

Of course, there are many judgments whose truth can't be determined just by analyzing the terms. In order to decide their truth, we have to appeal to something outside the terms themselves. In that case, well, what's the opposite of analytic? It's synthetic, so in that case the judgment in question is called a "synthetic judgment." Let me repeat: If we must appeal to something outside the terms of the judgment to decide on its truth, it's called a synthetic judgment.

Examples include: "The grass is green"—we've got to actually look. "The grass is green" is not true just by definition—we know that; if you don't water your grass, it turns brown. Or "A bachelor is unhappy"—maybe so, but you can't tell just by analyzing the term "bachelor." To find out if "A bachelor is unhappy" is a true judgment or a false judgment requires an empirical test—so it's a synthetic judgment.

Now let's focus more closely on synthetic judgments. To determine the truth of a synthetic judgment, I said: "You need to appeal to something beyond just the terms of the judgment." The examples I've given so far— "The grass is green"; "The bachelor is unhappy"—appeal to sense experience. Is that the only thing you can appeal to? Kant says: "You can appeal either to sense experience, or to something other than sense experience." Now it's time to introduce the next-to-last of the technical terms Kant uses to make this point precise.

If what we appeal to in determining the truth of a judgment is sense experience, the judgment is called an "*a posteriori* judgment." Kant says that any judgment whose truth we can decide on the basis of sense experience, or empirical investigation, is an *a posteriori* judgment. The Latin word *posterior* means "later"—so we're basically saying "after experience," "later than experience." All judgments involving empirical evidence are *a posteriori*. How do we know that honey is sweet or that the sky is blue? From the senses—so, "Honey is sweet"; "The sky is blue"—those are *a posteriori* judgments.

Last, but not least, a judgment whose truth can be determined independently of any particular sense experience by appealing to something other than sense experience, Kant calls an "*a priori* judgment." If what we appeal to in determining the truth of a judgment is independent of sense experience, the judgment is called an *a priori* judgment. The Latin *prior* means "before," so we determine the truth of a judgment *a priori* before observation or experiment.

Let's review Kant's terms. If the truth or judgment is determined just by analyzing the terms, it's an analytic judgment. A triangle has 3 angles, for instance. You analyze the term "triangle"—hey, you get 3 angles. If we appeal to something outside the terms of the judgment to decide on its truth, it's a synthetic judgment—like: "The grass is green." If what we appeal to is sense experience, it's an *a posteriori* judgment—"Honey is sweet." If what we appeal to is independent of sense experience, it's an *a priori* judgment—like: "All right angles are equal," maybe.

Now we have 2 sets of 2 categories. The multiplication principle says there are 4 possible combinations of the terms we've used. Here they are: Analytic *a priori*; Analytic *a posteriori*; Synthetic *a posteriori*; Synthetic *a priori*. Let's look at each of these 4 in turn, and see if there are any such judgments. First, analytic *a priori* judgments—these obviously exist. Any judgment that's true just by definition will qualify. Besides the standard philosopher's example: "All bachelors are unmarried," consider one we've seen before: "At least half of the test scores will be at or above the median." That judgment is "analytic" because we know that it's true just by analyzing the term "median"—it says exactly that half the scores are at or above it. We do not need to appeal to sense experience to tell that this judgment is true. Judgments like that often look as though they're giving us new information.

But if we know what the terms mean and if we analyze the terms, *analytic a priori* judgments are really just unpacking the definition of the terms—and they all, ultimately, reduce to things like: "George Washington's white horse was white."

In Lecture Twenty-Three, I told you about a criticism of Spinoza's proofs in his *Ethics Demonstrated in Geometrical Order* on exactly these grounds: Spinoza has packed the conclusions he wants to reach into his definitions—so when you unpack the definitions, the conclusions just fall out. You can decide if you agree with that criticism of Spinoza, but using Kant's language, what that criticism says is: "The judgments in Spinoza's *Ethics* are all analytic."

Anyway, I think it's clear now that analytic *a priori* judgments do exist. Analytic judgments are, by definition, *a priori*—independent of experience—because you don't need to know anything beyond the terms in the judgment to determine whether the judgment is true. Just analyze the terms.

Are there analytic *a posteriori* judgments? No. That's a contradiction in terms. By definition, you don't need to appeal to anything outside the terms to determine whether an analytic judgment is true or not—so, certainly, you don't need to appeal to sense experience.

Now we turn to synthetic judgments. Synthetic *a posteriori* judgments obviously exist. Any non-analytic judgment about the world of sense experience requires an appeal to something outside the definitions of the words involved—and that "something" we appeal to is the world of experience. There are literally millions of examples of synthetic *a posteriori* judgments besides: "The grass is green." How about: "Mars goes around the Sun in an orbit shaped like an ellipse"? Or "If you graph the heights of all the people in a population, the frequency curve will be bell-shaped"? Or "It's hot outside today"? These are all synthetic *a posteriori* judgments.

Now for the one that really matters to Kant. Is there anything in the fourth category? Are there any synthetic *a priori* judgments? Can synthetic *a priori* judgments exist? Let's see why this question matters so much to Kant. What Kant is after is to see if metaphysics is possible. So let's look at the question from the other side—from the side of metaphysics. Under which of Kant's headings should we classify the judgments of metaphysics?

Kant says that the judgments of metaphysics can't be *a posteriori*. I argued that earlier. Let me give yet another of my examples to illustrate Kant's general argument. Look at the statement: "Every effect has a cause." Someone objects and says: "Hey, wait, comets don't have a cause. They just appear out of the blue. There's no cause for them. They just happen."

Then along comes Edmond Halley and gives a causal explanation: "Comets are bodies that orbit the Sun in very long ellipses." Okay? So we can see them only occasionally, when they come close to the Earth. But do the many successes we've achieved like that mean that we'll always be able to find a causal explanation for every fact we see in the world? Because of the word "always," we can't test the statement: "Every effect has a cause" by doing experiments or by looking at the history of science. So things like: "Every effect has a cause" cannot be *a posteriori* judgments.

All right. So the judgments of metaphysics have to be *a priori*. Are they analytic, or are they synthetic? If the judgments of metaphysics are analytic, they've got to be true by definition. Are they? Do we define "the world of sense experience" as "following laws"? No, we don't. We just define the world of sense experience as what we perceive.

Do we define observed facts as things that must have causes? Again, no we don't. An observed fact is just something we observe. The truth or falsity of an analytic judgment depends only on the meaning of the words and the principles of logic—notably the principle of contradiction. We know that "A bachelor is married" is false just from that. But it isn't a contradiction to say: "We observe facts that don't have causes." So, for Kant, the judgments of metaphysics are not analytic.

There's now just one possibility left: The judgments of metaphysics must be synthetic *a priori*. So we're back now to our earlier question once again—but now it's clear why Kant cares about it: "Do synthetic *a priori* judgments exist?" If they exist we ought to be able to find them. So, Kant asks: "Are there any synthetic *a priori* judgments whose truth is universally recognized?"

Suppose there are. Suppose Kant could actually show you some synthetic *a priori* judgments, and then another question immediately arises: If there are synthetic *a priori* judgments, how is it that we come to know that they are true? Or as Kant puts the question: "How is synthetic *a priori* knowledge possible?" How can we have knowledge that is independent of experience, and yet not true just by analyzing the terms we use in the judgment?

"Metaphysics," Kant says, "stands or falls with the solution of this problem." So we need, first, to find some synthetic *a priori* judgments. Once we've found them, we need to explain how we know they're true. In the next lecture, I'll show you exactly how Kant does this. But right now, I will tell you where he finds them—where he finds a lot of synthetic *a priori* judgments. He finds them in Euclid's geometry.

In the next lecture, we will see how Kant—in his celebrated book, the *Critique of Pure Reason* (the date is 1781)—Kant both identifies geometric theorems as synthetic *a priori* judgments and explains how it is that we come to know that they are true. We'll also see some important and influential things that Kant says about space and time. The next lecture will also include some closely related ideas held by some of Kant's contemporaries and successors. But right now, let me repeat: Kant says that we can find synthetic *a priori* judgments—that is, judgments that are independent of experience and that aren't true just by analyzing the terms. Kant says we can find synthetic *a priori* judgments in Euclid's geometry.

Marking off mathematics as a special kind of knowledge like this immediately suggests that Kant has a lot in common with Plato—and he does. Plato and Kant both share the belief that statements about the world of sense experience are not the only kind of knowledge. There is knowledge that goes beyond sense experience. But there is one key difference between Plato and Kant.

Plato puts his forms or ideas in a transcendental realm. For Kant, the ideas are in the human mind. I'll talk about what Kant says about mathematics in detail next time. But before closing this lecture, we ought to look at a non-mathematical example of how Kant uses ideas in the human mind to make sense out of the world of sense experience. I want to see with you what Kant says about causality.

Kant says there are "things as they really are," but we can't possibly know things as they really are. What we experience, instead, is the world of appearances—how things appear to us. The Greek word for what appears to us is *phenomena.*

So, we see appearances—phenomena—that the real things as they are somehow produce. So our science isn't about things as they really are; our science is about the world of phenomena. What is it that lets us make laws out of the world of phenomena? It is the human intellect.

Here's an example. Suppose you take a cold stone and put it out in the Sun. The Sun shines onto the stone. We observe that. We touch the stone, and now the stone feels warm. We observe that. David Hume comes along and says: "Yes, first the Sun shines on the stone—then the stone gets warm"—2 empirical observations. That's the whole story. But Kant disagrees. He disagrees because we don't say what Hume said. We don't say: "First the Sun shines on the stone—then the stone gets warm." What we say is: "The Sun warms the stone." We've added an idea to these observations about the

Sun and the warmth of the stone—the idea of causality—when we say: "The Sun warms the stone."

Correlation is something we observe; causality is an intellectual category. Yet, Kant says: "We can't think about the world without it. We can't think about the world without thinking in terms of 'causality,' which is in the intellect—not the world. Causality is in the mind. The causality is not out there—but we can't think without it." That is Kant's answer to Hume.

Pursuing this matter farther would take us beyond the bounds of this course. But now we have enough of the Kantian machinery so that, in the next lecture, we'll be able to understand how Kant uses the example of geometry to show that synthetic *a priori* judgments do, in fact, exist—and, therefore, that metaphysics is possible. Thank you.

# Lecture Twenty-Eight
## Kant's Theory of Space and Time

**Scope:** In this lecture, we will see how geometry provides examples of the existence of the kind of statements—synthetic a priori statements—required by Kant's view of metaphysics. We will see that Kant, unlike Newton and the Newtonians, locates space in the mind, not in the outside world. How does Kant establish his view of the nature of space? We will see how Euclidean geometry is presupposed by him and ask whether Euclidean geometry is therefore the only one possible. We will raise, but not answer yet, the question of what Kant's views portend for the philosophy of mathematics, space, time, and human thought. Finally, we will see how other mathematicians and philosophers of the day also assumed the necessity of Euclidean space, though in different ways and for different reasons than Kant did.

## Outline

I. As we suggested in the previous lecture, Kant is another in the line of philosophers who have appealed to mathematics to address an important philosophical question.

   **A.** Kant wants to demonstrate that metaphysics is possible.

   **B.** He establishes that metaphysics must consist of synthetic a priori judgments.

   **C.** As we shall see, Kant says a great deal more, much of which is profound and influential, about geometry, space, and time.

II. As we have frequently observed, most mathematicians and philosophers are strongly convinced that geometry is not about the empirical world, so its judgments must be a priori, not a posteriori.

   **A.** The question, then, is whether they are analytic or synthetic.

   **B.** Imagining the judgments of geometry to be analytic is an attractive possibility, since after all, one begins by stating definitions, and we frequently analyze the definitions in the course of a proof.

**C.** But Kant says no, the judgments of geometry are synthetic, and he shows us this by carefully discussing how we come to know the truth of a particular proposition in geometry: that the sum of the angles of a triangle is 2 right angles.

**III.** Let us review how we proved that the sum of the angles of a triangle is 2 right angles.

   **A.** If this were an analytic judgment, we could have proved it by analyzing the term "sum of the angles of a triangle."

   **B.** But you can analyze this term as long as you wish, and what you will get is "3 angles" (because it is a triangle), "3 sides" (because it is a triangle), and "sum"—but with no way of breaking that down further into what the sum is.

   **C.** We turn instead to the way a practitioner of geometry approaches the problem.

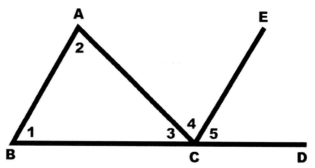

   1. The first step is to construct a triangle.
   2. The second step is to extend the base of the triangle.
   3. The third step is to divide the exterior angle just formed by constructing a line through the corner parallel to the opposite side of the triangle.
   4. Once these constructions are made, we argue about the equality of various angles, showing that the sum of the angles of the triangle add up to the sum of the angles along the straight line at the corner, which form 2 right angles.

   **D.** The essential feature of this proof is the construction.
   1. Where did we make the construction?
   2. Not on paper, says Kant; we made it in space.

**IV.** What, for Kant, is space?

    **A.** It is not empirical; it cannot be sensed.

    **B.** It is in our minds.

    **C.** We order our perceptions in space; we say, "This is above that" or "This is to the left of that."

    **D.** He observes that we can imagine a space without objects but that we cannot imagine objects without space.

    **E.** Space, for Kant, is the form of all possible perceptions.

    **F.** Space is a pure, unique, a priori intuition of the intellect.

    **G.** Notice that if space is unique, and if one thing that is true in it is that the sum of the angles of a triangle is 2 right angles, then space is Euclidean.

    **H.** Kant never imagined that anything other than Euclidean space was possible, so one cannot say that he denied that there could be a non-Euclidean space.

    **I.** Still, his philosophy was taken to mean what Voltaire had said for different reasons: Geometry had to be Euclidean.

**V.** Before we leave Kant, we ought to look at what he says about arithmetic and what he says about time.

    **A.** Not only is geometry synthetic a priori for Kant, so is arithmetic. Once again, Kant bases this conclusion on the process of construction.

    **B.** Kant considers time to be very much like space—a pure a priori intuition of the intellect.

**VI.** There were many other thinkers in Kant's time who argued that space had to be Euclidean, though their arguments were different from Kant's.

    **A.** Both the physics of René Descartes and the physics of Isaac Newton required space to be the same in all directions, so that a moving object not acted on by forces could continue to move at a constant speed in a Euclidean straight line forever.

    **B.** The resolution of forces using what is called "the parallelogram of forces," and therefore Euclid's theory of parallels, underlies all of classical physics.

    **C.** Leonhard Euler agreed with Newton that space had to be real. Space could not be in our minds, because physics—the science of the real world—depends on it.

**D.** Johann Lambert and Jean-Baptiste Fourier thought that the parallel postulate could be derived from the law of the lever in physics, so that geometry was a physical science but still Euclidean.

**E.** Joseph-Louis Lagrange tried to prove the parallel postulate by relying on the principle of sufficient reason. He never published the proof, however—perhaps because he realized that it relied on the same Euclidean space that he was trying to prove.

**F.** Pierre-Simon Laplace said that a consequence of the mathematical form of Newton's law of gravitation is that if the size of all bodies and distances and velocities in the whole universe were to decrease proportionally, the bodies would describe the same curves that they do now, so that universe would still look exactly the same.

    **1.** This means that all that matters is the ratios of distances, velocities, and so on—not their values.

    **2.** So, Laplace says, "the idea of space includes the following self-evident property: similar figures have proportional sides."

    **3.** This property of similar figures rests on Euclid's theory of parallels, so Laplace is saying that Newton's physics requires space to be Euclidean.

**G.** Even some of the most radical mathematical thinkers of this time, like Thomas Reid, William Hamilton, and Giordano Bruno, were caught up in this general sense that there was no alternative to Euclid's plane geometry.

**H.** Finally, as we will see in the next lecture, art and architecture contemporary with the philosophers and scientists just mentioned reflect the Euclidean nature of space.

## Essential Reading:

Kant, *Critique of Pure Reason*, introduction, sec. 1 of Transcendental Aesthetic ("Of Space").

## Suggested Reading:

Friedman, *Kant and the Exact Sciences*.

Grabiner, "Why Did Lagrange 'Prove' the Parallel Postulate?"

**Questions to Consider:**

1. Are the arguments that space has to be Euclidean just like the arguments that the world has to be flat, because the locality we live in looks like that?

2. What other possibilities are there for the nature of space besides Kant's idea that it is a pure a priori intuition of the mind?

# Lecture Twenty-Eight—Transcript
# Kant's Theory of Space and Time

Welcome back. Today's lecture has 2 main goals. The first goal is to show how Kant uses the example of Euclidian geometry to show that there are synthetic *a priori* judgments—judgments that are independent of experience, but whose truth can't be determined just by analyzing the meaning of the terms. The second goal of this lecture will be to go beyond Kant, and to show how ideas about the metaphysical necessity of Euclid's geometry permeated 18th-century thought in general. That's a lot. So let's get started.

As I said in the previous lecture, Kant wants to show that metaphysics is possible. He says that metaphysics has to be composed of synthetic *a priori* judgments—so he needs to show that there are synthetic *a priori* judgments, and to explain how we can tell whether they are true. In this lecture, we'll see how Kant does all of that. We'll also see that he says a great deal more—much of which is profound and all of which is influential—about geometry, space, and time.

So, together with Kant, let's look again at Euclid's geometry. Virtually all the mathematicians and philosophers that we've discussed so far have agreed that geometry is not about the empirical world. Euclid's lines that have no width, his circles that are perfectly round—those things don't exist in the physical universe. But don't we get the ideas of lines and circles from sense experience? Yes—psychologists might say: "You don't get the idea of a circle without being out in the world of experience and looking, say, at the Moon or the Sun. You don't get the idea of a straight line without looking at, say, a tree growing straight up." But that's a biographical story about how human beings developed those ideas.

But the ideas themselves, according to Kant, are independent of any particular experience—and we don't determine the truths about these ideas by sense experience. Furthermore, geometry is general. If we prove, say, that the sum of the angles of a triangle is 2 right angles, it's true for all triangles—not just the one I drew in the diagram of the proof. So the judgments of geometry, according to Kant, are *a priori*. *A priori*, okay—but are the judgments of geometry analytic or synthetic? Recall that an analytic judgment is one whose truth we can determine just by analyzing its terms, while a synthetic judgment requires an appeal to something outside of its terms.

At first, it seems as though the judgments of geometry should be analytic. After all, we start studying geometry with definitions. We often analyze the terms, using the definition, in the course of a proof. For instance, you'll recall that in proving Euclid's very first proposition, we had a circle. In the proof, we needed 2 of the radii of that circle to be equal. We concluded that the 2 radii were equal by analyzing the notion of "circle" into "curved line" and "every point on it having the same distance from the center." But in general on this question, Kant says "No."

Although there is some analytic reasoning in geometry—my circle example shows this—the judgments of geometry, the proved propositions according to Kant, are synthetic. Kant doesn't just tell us this—he shows us. Kant shows us this by explaining how we come to know the truth of a particular proposition in geometry. What proposition does he pick? "The sum of the angles of a triangle is 2 right angles." Following Kant's argument about how we come to know that this proposition is true requires understanding Euclid's proof. So the work you put into mastering Euclid's proof of that theorem will now have a real philosophical payoff.

Let's review how Euclid proved that the sum of the angles of a triangle is 2 right angles. First though, if this were an analytic judgment, we could have proved it by analyzing the terms in the statement: "The sum of the angles of a triangle is 2 right angles." So let's analyze the term of the judgment "sum of the angles of a triangle." What do you get when you analyze that? What do you get when you analyze the concept of a triangle? You get 3 angles (because it's a triangle); you get 3 sides (because it's a triangle). What do you get by analyzing the concept "sum"? That something is being added.

You could analyze the concept "3" and maybe understand that better. But you can analyze the phrase "sum of the angles of a triangle" until you're blue in the face, and you will not get the value of that sum—whether that sum is 2 right angles, or 10 right angles, or what. So how do we know it? How does Kant prove—how does Euclid prove—that the sum of the angles of a triangle is 2 right angles?

Kant follows Euclid. The first step is to construct a triangle. So, here we have one. The next, very important step is to extend the base of the triangle. The next step is to divide the exterior angle that we've just constructed by constructing a line through the vertex of that angle parallel to the opposite side of the triangle. Once Euclid made those constructions, he argued about the equality of various angles. He showed that the sum of the angles of the triangle add up to the sum of these 3 angles here around the right-hand corner of the triangle, which form 2 right angles.

But the essential feature of this proof, according to Kant, is that we made these constructions. The proof can't proceed without those constructions. The proof requires the extension of the base, and then requires the construction of that parallel to the side, which is what makes those angles over there on the right. That line and those angles didn't exist when we got started. We needed them in the proof, and we had to construct them.

Here is Kant's dynamite question: "Where did we make those constructions?" Certainly not on paper, says Kant. That's not what geometry is about. Not on your computer screen. As I have been saying throughout the course, the physical lines we draw are just imperfect pictures of the actual construction. So where did we make the construction?

Here's Kant's answer: "We made the construction in space." Let me say it again: "We made the construction in space." According to Kant, what is space? First—what it's not: Space is not empirical. Space is not an object of sense experience. You can't see space. You can't touch space. So far, he's with Newton. But where is space? Here Kant breaks with Newton. Space is not "out there" somewhere. Like causality, space, for Kant, is in our minds—always in our minds. We can't think about the world without it—without space.

Here are some of the things Kant says to convince us that space is in our minds. He says—and this is an empirical claim, so you should try it—he says: "We can imagine a space without objects, an empty space, but," he says, "we can't imagine objects without space."

What do we do with the "space" that's in our mind? What we do with it is we order our perceptions in space. We say: "This is above that." We say: "This is to the left of that," and so on. We use the space in our minds to put our sense perceptions in order. "Space," says Kant, "is the form of all possible perceptions." Let me say that again. "Space is the form of all possible perceptions."

To use just one more of Kant's technical terms—"intuition"—space is a pure *a priori* intuition of the intellect. For Kant, an intuition is something unique. So there's just one space in our minds. Now if the space in our minds is unique, and if the sum of the angles of a triangle in this space is 2 right angles, then the space in our minds is Euclidean—that is, Kant's space has to obey Euclid's fifth postulate.

Kant never imagined that anything other than Euclidean space—that is, a space in which Euclid's fifth postulate holds—was possible. Maybe if Kant had known about non-Euclidean geometry, he would have had fascinating and insightful things to say about it. But non-Euclidian geometry didn't exist in 1781. Kant didn't know about it, so he did not address that question at all.

It isn't really fair to say, as some philosophers say, that Kant denied that there could be a non-Euclidean space. The possibility doesn't seem to have occurred to him. Still, Kant's philosophy was taken by others to reaffirm what Voltaire had said earlier: "There's just one geometry. Geometry has to be Euclidean." This is necessary. Hold onto that thought; I'll get back to it later.

But now let us return to Kant's key question and answer it. His question was: "Are there any synthetic *a priori* judgments?" The answer is "Yes." Geometry is full of them. We don't establish the truth of theorems in geometry by analyzing the terms in them. We prove these theorems by appealing to something outside the words of the theorem. We appeal to our intuition of space, because we establish the truth of the theorems of geometry by making constructions in our intuition of space. So Kant has answered the question posed in the *Prolegomena*: "Yes, synthetic *a priori* judgments do exist—therefore, metaphysics is possible."

There's a lot more to Kant's philosophy, and we can't do it all. But before we leave Kant, we ought to look at what he says about arithmetic, and what he says about time.

First, arithmetic—not only is geometry synthetic *a priori* for Kant, so is arithmetic. That's not what Leibniz would have said. As I mentioned a few lectures ago, Leibniz wanted to invent a symbolic logic, so that 2 people who disagreed about something could resolve it just by calculation. That, Leibniz thought, would make every judgment into what Kant called an analytic judgment. You find out that it's true just by analyzing the terms. Leibniz gave a proof that 2 + 2 = 4 by a process that he saw as just analyzing the terms. It went more or less like this:

Is 2 + 2 = 4? Well, 2, by definition, is 1 + 1, so we break down 2 + 2 into (1+ 1) + (1 + 1). We can get rid of the grouping—so we do, and then we can re-group them. The first 1 + 1 we group together—we say, by definition, that's 2. So this line is 2 + 1 + 1. But 2 + 1—we group those, and by definition that's 3. So, this turns into 3 + 1. But 3 + 1—that's the definition of 4. So the proof succeeds by analyzing "2 + 2" into a sum of 1's, and then using that sum of 1's to construct the number 4.

But Kant didn't think that proofs like this are analytic. Kant says: "Look at what you're doing. When you put all those 1's together, what you are doing is you're constructing the number 4 out of the pieces of 2 + 2. That's a construction, and in the case of arithmetic, that construction is a process that takes place in time. Leibniz's proof works only because you carry out the steps in order in time. You've constructed the answer, and every step of this construction takes place in time.

All right, Professor Kant, what is time? Kant says time is very much like space. We order our perceptions in time. We say: "This one event is earlier; this other event is later." Time, then, like space, is a pure *a priori* intuition of the intellect—and time, like space, is unique. In this, as in many other conclusions, Kant's philosophy has been very influential. You can spend a lifetime studying Kant.

But other thinkers in the 17th and 18th centuries also weigh in on the nature of geometry and space. There's something new going on in this time period—there's a sense in which the whole subject matter of geometry has changed.

When we were talking about Plato, or Aristotle, or Euclid, we took it for granted that geometry was about circles, triangles, parallel lines—things like that. But in the 18th century, even before Kant and very much reinforced by Kant, people started talking about geometry as being about space. This new emphasis on space had many causes, but the key players were Descartes and Newton.

First, Descartes—Descartes thought that the essential attribute of matter was that it occupies space. Descartes also, as I said earlier, thought that space was full—full of ether, not empty. So geometry, for Descartes, was about a space full of matter. But this geometric space, for Descartes, had to be Euclidean. Of course, just like Kant, Descartes never imagined any other kind of space. But Descartes's space needs to be Euclidean, because the theory of parallel lines is essential for Descartes's analytic geometry.

As for Newton: Newton's physics, as we saw before, requires a real space, and a physical body with no forces acting on it moves in a straight line with respect to that space. Newton's space is infinite, and it's the same in every direction. Furthermore, the analysis of motion in Newton's physics is always done in terms of Euclidean geometry and is loaded with ideas from Euclid's theory of parallel lines. For instance, an important technique in Newton's physics is the use of what is called a "parallelogram of forces."

Here is the very first proposition proved by Newton in *The Principia*. Put a physical body here at A. Let 2 forces act on it—one in the direction of AB, and one in the direction of AC. If the physical body is acted on by those 2 forces simultaneously, where will the body go? It will go along the diagonal of this parallelogram in the same time that it would take to go along the sides of the parallelogram if it were acted on by those forces separately.

That's just one example of Newton's use of parallelograms of forces. What is important for us is that the mathematics of parallelograms he uses rests on Euclid's theory of parallels. So the space in which these bodies are moving—and remember, that for Newton these are real motions since they're produced by forces—the space these bodies move in is Euclidean. That is, Euclid's fifth postulate, and all the theorems that are based on it, must be true in Newton's space.

Many other thinkers in the 18th century argued that the space in which physical laws hold had to be Euclidean—had to be. These were among the leading philosophers and physicists of the age.

For instance, in the middle of the 18th century—that's earlier than Kant— the great mathematician and physicist Leonhard Euler wrote an essay called "Reflections on Space and Time." In this essay, Euler agreed with Newton that space had to be real. Space couldn't just be the relations between bodies the way Leibniz had thought.

Why? Because of Newton's first and second laws of motion, said Euler. These laws, Euler said, were beyond doubt because of their marvelous agreement with actual observations. Look at Newton's first law about the motion of a body with no forces acting on it. The straight-line, constant-speed motion of this single body can't possibly depend on where other physical bodies happen to be at this time. That would be Leibniz's idea, right? That doesn't make physical sense.

No, the straight-line motion has to be measured with respect to immovable space. Space isn't in our minds, says Euler. How can physics—the science of the real world—depend on something in our minds? Many thinkers gave a deep philosophical reason that space has to be Euclidean. There is this one idea—common to Descartes, Newton, Euler, and Kant—that space is the same in all directions. That idea, by the way, is presupposed every time we draw a diagram in geometry.

We usually choose to draw a triangle with its base on the bottom and a point at the top, but this makes no difference at all to what we're proving, and we could equally well do it the other way. The idea that space is the same in all directions is intimately related to Leibniz's principle of sufficient reason.

There's no reason to prefer one direction to the other—therefore, space is the same in all directions. The principle of sufficient reason was used a lot in the 18<sup>th</sup> century to argue that space is Euclidean. A standard example of the principle of sufficient reason, as I've mentioned before, is that a lever with equal weights at equal distances from the point of support must balance.

In the 18<sup>th</sup> century, that fact about levers is itself used to establish the Euclidean nature of space. For instance, the Swiss mathematician Johann Heinrich Lambert said consider the forces exerted by the weights at the end of a lever like this, with equal weights at equal distances from the fulcrum. Those equal forces have the same direction—downward—and the forces are parallel.

Because the lever stays on the point of support, the point of support is exerting an upward force on the lever that's equal to the downward force of the weights. That upward force is parallel to the downward forces of the weights. This is all about parallels. On the basis of these ideas, Lambert thought someday we could use a law of physics—the law of the lever—to prove the parallel postulate. Lambert's not the only one who thinks things like this: The French physicist Joseph Fourier also thought that the parallel postulate could be derived from the law of the lever in physics. But Fourier didn't think of the law of the lever as derived from a metaphysical principle, like sufficient reason.

Fourier thought the law of the lever was a physical law derived from experience. So getting the parallel postulate from the law of the lever would make geometry into a physical science. Still, though, the geometry would have to be Euclidean.

Another example: Joseph-Louis Lagrange. Lagrange was another great mathematical physicist and mathematician. Lagrange's major work on physics is full of appeals to symmetry and sufficient reason. Lagrange calls the balancing of a lever with equal weights at equal distances "a self-evident truth." It's self-evident, but Lagrange gives a reason for it anyway. Lagrange says: "It's a self-evident truth because there is no reason why either of the weights should move." But what is even more interesting is that Lagrange once actually tried to prove the parallel postulate—in the equivalent form of Playfair's axiom about the uniqueness of parallel lines. Lagrange actually tried to prove the postulate on the basis of the principle of sufficient reason. It was an indirect proof. Here's how he did it.

Here's a line outside a point on one parallel. Now suppose that there is another parallel to the original line through the point P. It might look like this. But, by the principle of sufficient reason, there is no reason that the new parallel line should be below the point P on the left, and above the point P on the right; it could equally well be drawn the other way. So the principle of sufficient reason says that there must be another symmetric one that goes the other way—as in this new diagram. Lagrange repeats the argument again for lines symmetric to his new parallels—so you get one above and one below the new parallel, and you keep doing this over and over again, and eventually you get something like this, which he says is absurd. So, he concludes, there can only be one parallel.

Lagrange never actually published this proof. Maybe that's because he realized that, by assuming that space is the same in all directions and that everything is symmetrical, he was assuming the very Euclidean nature of space that he was trying to prove. But the fact that a great mathematician like Lagrange linked space being Euclidean with the principle of sufficient reason—in geometry, as well as in physics—shows how closely allied all these ideas had become in the 18th century.

Let me give you one more example of a great physicist who thought that it was absolutely necessary, because of Newton's physics, that space had to be Euclidean: It's our friend Pierre-Simon Laplace. Laplace said—I can't give you the details here; just trust Laplace—that one consequence of the mathematical form of Newton's law of gravitation—that's that gravity varies as 1 divided by the square of the distance between the bodies involved—one consequence is that, if the size of all bodies, and distances, and velocities in the whole universe were to decrease in the same proportion, the bodies would describe curves of the same shapes that they do now—like elliptical orbits, and so on.

The universe would look exactly the same—though on a smaller scale. This means that all that matters for physics—according to this idea of Laplace—is the ratios of distances, velocities, and so on—not their actual values, their ratios. So, Laplace says: "The idea of space includes the following self-evident property: Similar figures have proportional sides." This property of similar figures—that they have proportional sides—of course, as we've seen, rests on Euclid's theory of parallels. So Laplace has a new argument, based on Newton's physics, that space must be Euclidean.

Even some of the most radical mathematical thinkers of this time were caught up in this general sense that there was no alternative to Euclid's plane geometry. For instance, the Scottish philosopher Thomas Reid went so far as to claim that the "visible space" that we construct for ourselves has a different set of rules than Euclidean geometry does. That sounds pretty radical, right? That visible space doesn't follow Euclid's rules? But Reid, nonetheless, spent a fair amount of time trying to prove Euclid's fifth postulate from the others, trying to "fix" Euclid's geometry.

Let me tell you about William Rowan Hamilton, Irish mathematician. Hamilton invented an algebra in which multiplication was not commutative— that is, in which A times B was not equal to B times A. That was very revolutionary. But on Euclid, Hamilton followed the standard conservative line. Hamilton wrote in 1837: "No candid and intelligent person can doubt the truth of the chief properties of parallel lines, as set forth by Euclid in his *Elements*, two thousand years ago." Hamilton said: "The doctrine involves no obscurity nor confusion of thought and leaves in the mind no reasonable ground for doubt."

There's one more exciting idea that got linked with Euclidean geometry: the idea of infinity. In the Renaissance, we see the first clear expression of the idea that space is infinite. The infinity of space was justified by what was effectively the principle of sufficient reason by Giordano Bruno in the year 1600. Bruno said: "Of course, the universe has to be infinite, because there is no reason to stop at any point." That radical idea of Bruno's soon became a commonplace. Newton accepts it without missing a beat. So the prevailing view in the 18th century is that space must be infinite, flat but not curved, the same in all directions, perfectly symmetrical, and Euclidean. So the parallel postulate must be true.

The geometric necessity of these things was linked to Newton's physics, and it was linked to universal metaphysical principles, like the principle of sufficient reason. Whether it's out there [gesture] or in here [point to head], geometry is about space. Space has to be Euclidean—case closed.

As we'll see in the next lecture, artists got in on this, too. The art and architecture contemporary with the philosophers, and scientists, and mathematicians I've been discussing also reflect the Euclidean nature of space. As I hope to show you in the next lecture, art and architecture reinforced the Euclidean intuition of space in the minds of everyone who looked at the paintings, and who lived and worked in the buildings and public squares. From the Renaissance well into the 19th century, everybody saw space as Euclidean.

# Lecture Twenty-Nine
## Euclidean Space, Perspective, and Art

**Scope:** This lecture looks at the way our ideas of space were shaped by Euclidean geometry. We will describe the revival of geometric optics in the Renaissance using Euclid's optics and ideas about shape. Then we will look at the way geometry was used by painters and architects, including Piero della Francesca, Leonardo da Vinci, Albrecht Dürer, Michelangelo, and Raphael to map 3-dimensional space onto flat surfaces and to design buildings embodying geometrical balance and symmetry. We will conclude by describing the ways in which the Euclidean nature of space fits with philosophy and science in the period between the Renaissance and the discovery of non-Euclidean geometry in the 19th century.

## Outline

I. We start with the first real perspective painting, Masaccio's *Trinity*.

    **A.** We are used to pictures that look 3-dimensional, but people in the Renaissance were not, so this was very exciting to them.

    **B.** It is all done by means of geometry.

    **C.** Medieval examples show how much things had changed; the medieval works are carefully planned and organized, but there is no convincing 3-dimensionality.

II. Renaissance culture helped produce the new approach to the visual arts.

    **A.** Greek learning was revived.

    **B.** The invention of printing led to the rise of printed books—and pictures.

    **C.** The use of perspective developed in painting.

III. Euclid's *Optics* includes axioms and theorems that help us understand the geometry of perspective.

    **A.** Like his *Elements*, Euclid's *Optics* begins with a number of basic assumptions. Two are especially important.

        **1.** Assumption 2: The figure contained by a set of visual rays is a cone, with the vertex at the eye and the base at the surface of the object seen.

2. Assumption 4: Things seen inside a larger angle appear larger, those inside a smaller angle appear smaller, and those inside equal angles appear equal.

**B.** Proposition 6 of the *Optics* is a crucial one for perspective. It states that parallel lines, when seen from a distance, appear to be an unequal distance apart.

**C.** Euclid's *Elements* also has theorems that explain the properties of parallel lines and the proportionality of the sides of same-shape geometric figures.

**D.** Practical geometry includes finding the volumes of barrels; considering individual barrels as idealized geometric shapes helped people learn to "read" paintings.

**E.** Here are a few key facts about the geometry of perspective that allow us to "see" 3-dimensional reality on a flat canvas.
1. All parallel horizontal lines in the real scene that are perpendicular to the plane of the canvas are drawn on the canvas to meet at a point, called the "principal vanishing point" V.
2. Any set of parallel horizontal lines in the real scene that meet the picture plane at some angle are drawn to converge to a point, called a "diagonal vanishing point," somewhere on the horizon line Z.
3. Parallel horizontal and vertical lines in the real scene that are parallel to the picture plane are drawn, respectively, horizontal and vertical—and parallel.

**F.** Leon Battista Alberti was the first to write about perspective, as part of his treatise on painting.

**G.** Piero della Francesca wrote the first treatise devoted to perspective, explaining not only the how of perspective constructions, but also the why. He used these ideas in paintings of religious scenes.

**H.** Leonardo da Vinci used proportion and geometry in his paintings and drawings, for example *The Last Supper*.

**I.** Would one have to have mastered mathematical theory to create scenes like these? As it turns out, no. A drawing by Albrecht Dürer of how to draw a lute illustrates the practical technique. Dürer wrote the first book on geometry in German, with important material on perspective.

**J.** Dürer's student, Erhard Schön, published a number of pictures to show how to draw in perspective.

**IV.** We give more examples of important works of art that illustrate the use of the geometry of perspective.

    **A.** Raphael's *School of Athens* is a beautiful example of geometrical space, especially in its use of the principal vanishing point.

    **B.** Francesco Algarotti, the Italian popularizer of Isaac Newton, commissioned Tiepolo's *Banquet of Cleopatra*, which also shows a mastery of perspective.

    **C.** Canaletto's *Rialto, Venice, with Palladian Bridge and Palaces* is an instructive combination of architect Andrea Palladio's proportional designs, Canaletto's imagination, and Canaletto's mastery of perspective and sense of space.

**V.** We state some implications of the perspective revolution in painting.

    **A.** The viewer is in the same space as the people and objects in the painting.

    **B.** The objects in the world, even of religious subjects, are portrayed as an observer would actually see them.

    **C.** The material world has a rational order describable in terms of geometric forms.

**VI.** Architectural examples show the use of parallels and symmetry and also illustrate the shaping of the space people actually move through.

    **A.** Palladio's Teatro Olimpico in Vicenza, Italy, is a wonderful example of the use of parallels and symmetry to create an impressive unified space.

    **B.** Borromini uses perspective to make a corridor seem longer.

    **C.** Michelangelo's Campidoglio in Rome is our last and most magnificent example.

**VII.** We look at what later might look like non-Euclidean pictures and see how Euclidean even these are. We contrast these with the work of Escher in the 20th century, who was consciously portraying an alternative version of reality.

**VIII.** We conclude that scientists and artists long shared the desire to construct a model of the world as it appears to a rational, objective observer.

    **A.** Well into the 19th century, both scientists and artists constructed their models in Euclidean space.

**B.** Only in the 19th century did some mathematicians stop trying to buttress Euclidean geometry and begin to challenge it.

### Essential Reading:
Kline, *Mathematics in Western Culture*, chaps. 10–11.

### Suggested Reading:
Andersen, *The Geometry of an Art.*
Field, *The Invention of Infinity.*
Kemp, *The Science of Art.*

### Questions to Consider:
1. Find some reproductions of paintings in perspective that you like; if online, print them out, and if in a book, photocopy them. Find where the lines apparently perpendicular to the plane of the picture intersect—the principal vanishing point—by drawing these lines on your copy. Then ask why the artist chose to locate the principal vanishing point there.

2. Look at some of your favorite paintings, from the Renaissance through the 19th century, and at photographs of some major architectural achievements. Search for right angles, for parallel lines that are everywhere equidistant, for spaces that are bounded by parallel planes, and the like. Look at the room you are in right now for the same things. Do you agree with my contention that these experiences reinforce the idea that we live in a Euclidean, 3-dimensional space?

# Lecture Twenty-Nine—Transcript
# Euclidean Space, Perspective, and Art

Welcome back. This lecture will give a visual dimension to the philosophy that we've been doing in the past lectures—to look—literally, to look—at Euclidean space, perspective, and art. To do this, we'll begin with the Renaissance. Let's start with the first important perspective painting. That's the *Trinity* by Masaccio. The date of this is about 1428.

We are used to pictures that look 3-dimensional—because, after all, we have photography, and we have television, and we have DVDs. But, in the Renaissance, they didn't. So a painting like this was really exciting to them. A man of the Renaissance, the art historian Giorgio Vasari, said of this painting of Masaccio (it's actually a fresco; it's painted on a flat church wall in Florence), Vasari said: "It seems as though the wall is pierced."

This realistic illusion of depth, the feeling that the objects portrayed have pierced the wall, that feeling of depth that blew Vasari away—in Renaissance art, that feeling comes from geometry.

It's useful to look at a few medieval works of art so we can appreciate the difference between what Renaissance painters do and what their predecessors did. First, here's a picture from the *Bayeux Tapestry* from the 11th century. It's kind of charming, and it happens to show what we now call "Halley's Comet." It shows the comet appearing just before William the Conqueror beat Harold Godwinson at the Battle of Hastings. That's 1066, that battle.

Second, here's a charming agricultural scene from the 12th century. It's the month of September; it's from Spain. As you see, he's harvesting some crop.

Third, here's an action scene by one Nicholas of Verdun. This portrays the crossing of the Red Sea. This is from the biblical book, Exodus. The date of this portrayal is about 1181.

So, there are some examples. Medieval works of art can be religious or secular. They can be profound or playful. It's wonderful art, but there is no convincing 3-dimensionality.

In the Renaissance, though, we are in a different kind of space. So, here's sort of an outdoor example—this one maybe by Piero della Francesca, or as other scholars say from Luciano Laurana. It really doesn't matter. This picture is called *The Ideal City*. That is kind of a Platonic title, isn't it? It's from around 1470. Yes, this is a different kind of space. Look at it.

All right, this will do as an introduction. In today's lecture, I have 2 goals. First, I want to show how the geometry of perspective literally changed the way people look at the world. Artists used the geometry of perspective to make pictures look "real"—real in the sense of appearing to have 3 dimensions.

Second, I want to argue that Renaissance art, by doing these things that I said, that Renaissance art helped people internalize the idea that space is Euclidean. By "Euclidean," I mean such things as: Space is flat, not curved; parallel lines are everywhere the same distance apart, everywhere equidistant. Experiencing such a space in works of art was new in the Renaissance. I want to argue that this experience affected the way people actually see.

The use of perspective in painting arises from the Renaissance world-view in general, and from a revival of interest in geometry in particular. Starting in the 1400s, the Renaissance revived Greek and Latin culture in Europe— that's why we have the word "Renaissance," which means "rebirth"—but it isn't just a renewal of the old. The Renaissance also was full of things that were new, and novel, and exciting.

Here's a partial list. Number 1: The voyages of discovery that opened up the Americas to Europe; a signature date that's easy to remember is 1492. Second: An increased individualism and an emphasis on individual achievement—painters sign their paintings regularly, for instance. Number 3: The invention of printing and the rise of printed books—books, by the way, that included pictures, so you could reproduce drawings as well as texts. Number 4: Breakthroughs in science, like the work of Copernicus. Number 5: The rise of a money economy and modern banking; that's why we have all these words about banking that are from Italy—*banco*, it means "bench," "bank." Number 6: The replacement of Roman numerals by the Hindu-Arabic numbers—those are the ones that we use today. They make computation much faster, especially multiplication.

Last, but not least, the use of perspective in painting: The revival of Greek learning and the invention of printing worked together. In the Renaissance, there now are many copies available of Greek mathematical works and Greek philosophical works—like the *Elements* of Euclid, which was first printed in Venice in 1482. Euclid also wrote a book on

*Optics*. Both of these works by Euclid were essential for the development of perspective in painting.

I promise, we're going to look at a lot of art. But before we look at the paintings, we need to look at Euclid's *Optics*. Like Euclid's *Elements*, Euclid's *Optics* begins with a number of basic assumptions. Two of those assumptions are especially important. Here is the first of them:

I again give the number that it has there, even though I'm only going to do 2 of them. This is his Assumption 2: "The figure contained by a set of visual rays is a cone, with the vertex at the eye, and the base at the surface of the object seen."

Here's a modern illustration of the cone of vision out there on a football field. Also, some people call it a "pyramid of vision"—that's what Leonardo da Vinci called it. That's illustrated here. As you can see, we view objects in the angles that they form in the cone or pyramid of vision.

Now that we know that, we're ready for the other key assumption in Euclid's *Optics*: "Things seen inside a larger angle appear larger, those seen inside a smaller angle appear smaller, and those seen inside equal angles appear equal." You can see this right now. How do you know that this disk is smaller than this disk? The way you know is it makes a smaller angle with your eye; a cone of vision has a smaller angle. This fact about the angles is essential when you represent 3-dimensional reality on the 2-dimensional plane of a picture.

Let's look at one important proposition that Euclid proves in the *Optics*. This proposition—I'm not going to prove it, but I will state it—states that "Parallel lines, when seen from a distance, appear to be an unequal distance apart." We moderns can illustrate this really well with railroad tracks. There's a statement of his proposition, and there are the parallel lines.

So, from these examples, you can see the kind of thing that Euclid's *Optics* is about—and you can certainly see the importance of parallel lines and of visual angles.

There are also some results from the *Elements* of Euclid that are important in Renaissance art. I think that you will recognize them.

The first one is: "When 2 parallel lines are cut by a third line, the alternate interior angles [that's our friends, angle 1 and angle 2] and the corresponding angles [angle 3 and angle 2] are equal."

Then, from Book 6 of the *Elements*, we have this one: Given a triangle, remember if we draw the line DE parallel to the base, the line DE hits the sides of the triangle—then the little triangle ADE and the big triangle ABC are similar. They have the same shape, and then we have proportionality: AB is to AD as AC is to AE. The proof of that proposition, as you'll recall, ultimately rests on Euclid's Postulate 5.

Artists don't need to be professional mathematicians to be able to handle these geometric ideas. Actually, lots of people in the Renaissance had to be familiar with geometry in order to solve practical problems. For instance, wine and beer merchants needed to be able to estimate the volumes of the barrels that they shipped their products in. How do you estimate volumes of barrels? You treat real wooden barrels as though they were perfect geometric shapes.

The artist Piero della Francesca actually wrote a handbook about how to find the volumes of barrels geometrically. Getting a sense of what enclosed spatial volumes are like—for instance, volumes like the inside of a barrel—helps people learn to decode images in space that are bounded by curved surfaces—like, say, the dome of a cathedral, or, like a half-cylinder used as a ceiling. In fact, a ceiling shaped like the one in Masaccio's painting is still called a "barrel vault."

So we have the geometry we need for perspective, and we have a sense of what enclosed spaces look like. What else do we need? We need Plato. In the Renaissance, there was a renewed interest in the philosophy of Plato.

Plato's ideas helped encourage the view that mathematics underlies reality. So, geometry ought to underlie the representation of reality by artists. So, it's no accident that Leonardo da Vinci, echoing the motto of Plato's Academy, wrote: "Let no one read me who is not a mathematician."

Let us now turn to how all this geometry lets us see 3-dimensional reality on a flat canvas. This picture is the key to the whole thing. It's called the *Alberti Construction* after Leon Battista Alberti. I'll tell you a little bit more about Alberti in a minute. But for now, I want to concentrate on the geometry—this is a 2-dimensional drawing of a square floor that's divided into smaller squares—like a checkerboard. The floor is perpendicular to the plane of the picture, just like this checkerboard is perpendicular to the plane of your screen—these [horizontal] lines [are] parallel; these [vertical] lines [are] parallel.

Here are the rules for doing the drawing according to Alberti. Rule Number 1: "All parallel horizontal lines in the real scene that are perpendicular to the plane of the canvas are drawn on the canvas so that they meet at a point [See these guys all here? They all meet at this point V], and that point is called the 'principal vanishing point V.'"

Second: "Any set of parallel lines that are in the real scene that meet the picture plane at some angle [like these] are drawn to converge at a point like Z, which is called a 'diagonal vanishing point.'" See these lines that bisect the squares of the checkerboard, they are in the real world parallel, and they meet at this point, Z.

Third: "Parallel horizontal and vertical lines in the real scene that are parallel to the picture plane are drawn, respectively, horizontal and vertical—and parallel." In the case of the checkerboard, you see these guys are parallel to each other, and they're parallel in the picture.

These rules were all spellcd out by Alberti. Alberti lived in the 15$^{th}$ century. He was the first person to write a treatise on painting that includes material about the geometry of perspective. He described, in words (but without the picture) how to make the construction that I just showed you—this checkerboard construction.

Then, Piero della Francesca wrote the first treatise—beautifully illustrated, by the way—solely devoted to perspective in painting. Piero undertook to explain not only the how of perspective constructions, but also the why. That sounds like Plato or Aristotle, doesn't it? He knows that—Piero says: "Perspective is a true science."

The checkerboard construction we've just looked at was a model used by many painters. Renaissance artists often used squared-off pavements or other rectangular architectural features to emphasize the perspective in their paintings. Conversely, we can use the artists' representation of these parallel lines to locate the principal vanishing point. The eye of the painter—when painting the picture, and therefore the best position for the viewer's eye—is directly opposite the principal vanishing point.

Let's see an example of how Piero della Francesca himself used perspective, and these guiding lines, in a painting. This is Piero's painting, *The Flagellation of Christ*. So, look first at the ceiling, and the floor, and the up and down columns. They're parallel in the real world and parallel to the canvas to the plane of the picture—and, therefore, portrayed as parallel on the canvas. Where is the principal vanishing point? Let's look at where the things that in the picture are perpendicular to the plane of the picture—let's look at where they meet. Okay, so look at this brown line here (goes here). Look at this one (goes here). Look at this one (goes here). Looks like the principal vanishing point is on the principal actor, this guy with the whip.

Next, let's look at a painting that is very familiar to many people: *The Last Supper* by Leonardo da Vinci. It's rather hard to imagine that a building like this is where the Last Supper actually took place in the 1$^{st}$ century, but this is a Renaissance building.

Notice how the parallel lines in the ceiling, parallel lines in the walls, the ones that are parallel to the picture plane are sideways and up and down as they are in real life; and the ones that are perpendicular to the picture plane—the parallel lines in the actual scene perpendicular to the picture plane—converge. Where do they converge? They converge to the principal vanishing point, which is the face of Jesus. Obviously that says something about what Leonardo wants to emphasize in this painting. Leonardo was, of course, a consummate mathematician.

How could an artist construct scenes like this correctly? Did you have to have mastered all that mathematical theory? No, it turns out, you didn't. Here, for instance, is the great German artist and mathematician Albrecht Dürer showing us how to draw the musical instrument called the lute. What you do is here's the lute, and here's you—here's your eye.

You draw the lines of sight from the points on the objects through a screen—here's the screen, the screen is in the plane where the picture is going to be—to your eye. So you could construct the way a 3-dimensional object would appear on that plane—construct it one point at a time. So you see how to draw a lute in perspective, even if you don't know the geometry.

Dürer himself understood the theory as well as the practice. He wrote—and illustrated—a superb account of perspective in what was, in fact, the first book on geometry ever written in German. That was in 1525. So the Italian ideas about perspective were spreading northward through Europe.

Dürer had a student named Erhard Schön. Schön published a number of pictures to show you how to draw in perspective. Here's one that I think is especially charming: It's called *Placing Models on a Perspective Grid*. There were no complicated instructions. What Schön said in his book is: "Just look at my drawing—then you'll know what to do."

A German nobleman, one Count Johann von Simmern produced more such instructive drawings. Here's how he shows how to construct the heights of people. You see back here the principal vanishing point. It's pretty clear from the ceiling and floor where that principal vanishing point is. Look at the heights and the triangles formed by the converging parallel lines. You've got a triangle here. There's the vertex. There's the base, this guy. Here's the base of a smaller triangle, another guy. Count Johann's picture literally embodies Euclid's theorem that, in similar triangles, the sides are proportional. Here the heights of the guys are proportional.

Moving on from Germany into the Netherlands, here are 2 more instructional drawings. They're by the Dutch writer Johan Vredeman de Vries. This is the first one. It's another example of how to draw the guidelines to determine the heights of people in a drawing. Here's the second one. This is looking down 5 galleries from a balcony. De Vries has marked the lines that converge to the principal vanishing point there. If you put your eye at the right distance from the screen directly opposite the principal vanishing point, you'll feel like you are going to fall 5 stories. I don't know if I should encourage you to try this, but it's a really wonderful drawing.

That's a lot of geometry. Let's look at a few more works of art that this geometry helped to build. You'll remember from the first lecture, Raphael's *The School of Athens*. I've got to tell you when I saw this painting in the Vatican, the guide had to drag me out of the room. I just thought it was so neat. Look at the space that Raphael creates. Also, look for the principal vanishing point by following these parallel lines that are perpendicular to the scene. It's right here in the sky, between Plato and Aristotle.

Mathematically trained people, even if they weren't artists, were very interested in perspective in painting. This holds true not just in the Renaissance, but well into the 18<sup>th</sup> century. So the next example I want to show you is a painting commissioned by Francesco Algarotti. Algarotti was famous in Italy as a follower of Newton and as a popularizer of Newton's work. Algarotti actually wrote a book called *Newtonianism for the Ladies*.

The painting that Algarotti commissioned is by Tiepolo It's called *The Banquet of Cleopatra*, dated from the 1740s. You can see as you look at this painting both Tiepolo's mastery of perspective and his expansive sense of space. Algarotti also liked the works of Canaletto. Here's a Canaletto example.

Canaletto is famous for his paintings of large-scale scenic views, especially in his native Venice. This picture is supposed to be the Grand Canal at the Rialto in Venice, but not the way it actually was. To make Algarotti happy, Canaletto put in 2 extra buildings by the architect Andrea Palladio—the buildings were actually in Vicenza in northern Italy.

Canaletto also included this bridge. The bridge is based on a plan by Palladio, the architect—never carried out—for a new bridge at the Rialto. So this painting is an instructive combination of Palladio's proportional designs, Canaletto's imagination—and, again, Canaletto's mastery of perspective and his sense of space.

Let's stop and reflect for a minute. What has changed since the Middle Ages? What does seeing all this art do to our sense of space? Looking at these 2 works of art together, here is what I would say is new. Since the Renaissance, the viewer is in the same space as the people and the objects in the painting. You can imagine this floor just extending out to where you are. The objects in the world are portrayed as an objective observer would actually see them. It's a 3-dimensional perspective. The material world is given a rational order, and that rational order is described by means of Euclidean geometry. Now, one additional set of examples, and these examples will come from architecture. You see, architecture also uses parallels and symmetry. But there's more to it, because architecture shapes the space that people actually move through. We're going to look at 3 examples to illustrate how architecture shapes our space.

First, here again is the architect Andrea Palladio. This is his Olympic Theater in the city of Vicenza, completed in 1585 after his death. Palladio's building is a landmark—not only in the history of architecture, but also in the history of the theater. For our purposes, it's a wonderful example of the use of parallels and symmetry to create an impressive unified space. It takes the stage of the actors (over here) and the seating area of the spectators and brings them together.

For my second example, this is a corridor designed by Francesco Borromini. It's in the Palazzo Spada in Rome. Looks like a really long corridor, doesn't it? But, in fact, Borromini used a perspective trick to make this corridor appear longer than it really is. Look at a drawing of that corridor viewed from the top. You can see he puts the columns closer together as you go farther back. The sides of the corridor aren't really parallel at all; the sides converge towards each other. Let's look back at the corridor again. The person walking into this corridor is used to seeing architecture and painting in conventional geometric perspective. Similar figures, proportional sides—it's the 17th century; this has been around for a while. So the viewer sees these angles and, therefore, assumes that the end of the corridor is farther away than it, in fact, is. Let's look at the drawing again. The viewer, from the angles the viewer sees, thinks that the ends of the corridor are back here at A prime and B prime, rather than at A and B, where they really are—very impressive perspective illusion.

My last and most magnificent example is the city square—*The Campidoglio*—in Rome by Michelangelo. First, let's look at a 16th-century engraving of *The Campidoglio* by Etienne Dupérac. The date of this is about 1569. It's drawn from what is said to be the best viewpoint for this city square designed by Michelangelo. Second, here's a modern photograph of the Campidoglio. It's from the opposite direction from that of Dupérac's drawing. The space looks a lot deeper in this photograph. Why is that? Let's look again, as we did with Borromini, at the actual shape, viewed from the top of the Campidoglio. You can see here that the space in the so-called city square isn't a rectangle at all; it's a trapezoid. The fronts of these buildings on the side aren't really parallel. They diverge towards the palace at the back. So this city square is a trapezoid. Let's look at the photograph again. The convergence of those buildings—like the convergence of the columns in Borromini's corridor—the convergence of the buildings makes the open space appear more impressively deep. From the other side, the divergence of the buildings makes the palace at the back of the square look larger and even more impressive than it actually is. The large open area and the buildings together create a beautiful public space—a space that is, of course, Euclidean.

At this point, you might be thinking: "You've picked out only examples that prove your point." You might even want to object: "Come on! Is everything from the Renaissance to the 19th-century totally Euclidean in art? Didn't people ever imagine curved spaces?" That would be a very good question—because even in the Renaissance, some painters portrayed what we would now recognize as 3-dimensional curved spaces, using reflections in a convex mirror.

The most famous example is the *Arnolfini Wedding*, painted in about 1434 by Jan van Eyck. First, let's look together at the whole painting. I see all the Renaissance perspective stuff. Here's the convex mirror. Let's focus a little more closely on what you see in the convex mirror. To us modern viewers, the space portrayed in this convex mirror seems curved. Look at the sides of the windows: supposedly parallel, but not everywhere equidistant.

Ninety years later than this one, there's another reasonably well-known example—a painting by Parmigianino. It's his *Self Portrait in a Convex Mirror*. Again, the parallel lines aren't everywhere equidistant, and the space pictured seems curved. Look at the window at the top. Look at the ceiling. If you look at Parmigianino's hand, you might even say the space in this picture is distorted. A modern physicist, John Barrow—a really first-class physicist—has said that if people had just paid more attention to what things look like in convex mirrors, non-Euclidean geometry might have been discovered much sooner. But I am not so sure. I think instead that those few artists who painted scenes reflected in convex mirrors thought that what they were doing was solving an especially difficult problem in portraying 3-dimensional objects on a 2-dimensional Euclidean canvas. I don't think that these artists thought that they were creating an alternative kind of reality. In the 20th century, though, people who do this do think they're creating an alternative kind of reality when they use unconventional geometries. Here is one well-known example. M. C. Escher is well-known for creating alternative realities—and for him, this is one.

But for painting in perspective before the 20th century, I agree with the Oxford art historian Martin Kemp. Kemp says that from the Renaissance to the 19th century, the goal of constructing a "model of the world as it appears to a rational, objective observer" was the goal shared by scientists and artists alike." Let me repeat that: Scientists and artists alike constructed models of the world as it appears to a rational and objective observer. I would add that, virtually unanimously, artists armed with Euclid's *Optics* and Euclid's *Elements* taught us to see that rational world as Euclidean.

In the 19th century, a few mathematicians stopped trying to shore up Euclidean geometry, and instead began to challenge it. We will see how they did this in the next lecture. Later on, we'll see the implications of alternative geometries for philosophy, for science, and for art. But as a conclusion to this lecture, what I think is remarkable is that the people we're going to discuss in the next lecture, the people who invented non-Euclidian geometry—those people, in spite of all this Euclidean brainwashing from philosophers, from scientists, and from artists—what is remarkable is that those people were still able to imagine that the world could be otherwise. Thank you.

# Lecture Thirty
# Non-Euclidean Geometry—History and Examples

**Scope:** We are now ready to look at non-Euclidean geometry. We begin by looking at it from a logical point of view: What follows if we assume Postulate 5 to be false? We trace the history of attempts to prove the fifth postulate, especially those using proof by contradiction. We then describe the $19^{th}$-century realization, independently by Carl Friedrich Gauss, János Bolyai, and Nikolai Ivanovich Lobachevsky, that the supposed contradictory conclusions from the denial of Postulate 5 are valid theorems in a non-Euclidean geometry. We will also show physical examples of a concrete model for a non-Euclidean surface. Then, appealing to the logical structure of Euclid's theory of parallels, we will show that the theorems that should remain true even if Postulate 5 is false are in fact true on that non-Euclidean surface. And we see— literally see—that the theorems whose proofs required the truth of Postulate 5 need no longer hold on that surface.

## Outline

I. As we have said, for a long time many mathematicians had tried to prove Euclid's fifth postulate from the others.

   A. Questions about the fifth postulate began in the Greek-speaking world.

   B. Greek mathematics faded at the end of the Roman Empire but lived on in the Islamic world.

   1. Their mathematics was a synthesis of the Greek proof-based tradition with the computational traditions of the East, including China, India, Persia, and the Arabic-speaking world.

   2. An important example is the subject we still call by the Arabic name "algebra," in which the computation of solutions of verbally stated equations went along with geometric proofs that the solutions had the appropriate character.

   3. The mathematicians in the Islamic world who worked on Euclid's fifth postulate were among the major figures, including Omar Khayyam, Nasir Eddin al-Tusi, and Sadr al-Din al-Tusi.

**C.** In the 18<sup>th</sup> century, Italian mathematician Gerolamo Saccheri tried to use the technique of indirect proof to prove Postulate 5.

    **1.** He assumed Postulate 5 to be false, so if 2 lines are intersected by a third and the 2 interior angles on the same side of the third line add up to less than 2 right angles, the lines need not meet.

    **2.** From this assumption, logic led him to "absurd" conclusions.

**D.** What were some of these conclusions?

    **1.** Through an outside point, more than one parallel to a given line may be drawn.

    **2.** If you have a quadrilateral (4-sided figure) where 3 of the angles are right angles, the fourth need not be a right angle.

    **3.** The sum of the angles of a triangle is less than 2 right angles.

    **4.** Parallel lines are not everywhere equidistant.

**E.** But the "absurdity" of these conclusions turns out to be their disagreement with the standard ideas of geometry and space, rather than their being logically contradictory.

**F.** These conclusions seem contrary to our intuition of space.

**II.** In the early 19<sup>th</sup> century, 3 people, more or less independently, came to the same realization.

**A.** These conclusions are not contradictory or illogical or "absurd"; they are perfectly valid theorems—in a different geometry.

**B.** The 3 actors were Carl Friedrich Gauss, János Bolyai, and Nikolai Ivanovich Lobachevsky, working, respectively, in Göttingen, Germany; Hungary; and Russia, all outside the mathematical mainstream.

    **1.** Emphasizing the logic, Gauss called the subject "non-Euclidean geometry."

    **2.** Saying "I have created a new world out of nothing," Bolyai called what he had discovered "the absolute science of space."

    **3.** Lobachevsky called it "imaginary geometry," by analogy with imaginary numbers. Later mathematicians often call the geometry that denies Euclid's fifth postulate "Lobachevskian geometry."

**III.** Could a non-Euclidean surface actually exist? If so, what would such a surface actually look like?

**A.** It is often described as saddle-shaped. A physical model of a finite piece of this surface is called a hyperbolic paraboloid.

**B.** Since it is not a flat Euclidean plane, we need to redefine "straight line."

**C.** We do this by defining a straight line between 2 points, on any given surface, to be the shortest distance on that surface connecting those 2 points.

**IV.** We now turn to a truly exciting visual demonstration.

    **A.** On one Lobachevskian surface, where the lines drawn are all shortest distances, we can settle the truth status of all the propositions in Euclid's theory of parallels.

        **1.** On this surface, Postulate 5 is false.

        **2.** Propositions 27 and 28 are true; that is, if the relevant angles arc cqual, the appropriate lines on the surface will never meet.

        **3.** Proposition 29 is false on this surface; if 2 lines are parallel—that is, never meet on the surface—we see that the alternate interior or corresponding angles need not be equal.

        **4.** Proposition 30 is false on this surface, since we can see 2 lines through a given point, neither of which intersects a third line.

    **B.** We respond to arguments that this is not truly a surface.

**V.** We raise, but do not yet answer, some questions that all these results suggest.

    **A.** Could there be a geometry that denies one of the other postulates of Euclid?

    **B.** We have shown that there is a 2-dimensional, non-Euclidean geometry; could there be a 3-dimensional, non-Euclidean geometry?

    **C.** Does non-Euclidean geometry have any practical value, or is it just an intellectual curiosity?

## Essential Reading:

Gamow, *One, Two, Three—Infinity*.

Gray, *Ideas of Space*, 106–128.

## Suggested Reading:

Bonola, *Non-Euclidean Geometry*.

Katz, *A History of Mathematics*.

Petit, *Here's Looking at Euclid*.

**Questions to Consider:**

1. What would Euclid say about all of this?

2. Nikolai Ivanovich Lobachevsky wondered whether real space was Euclidean or not. With the methods available in the early 19<sup>th</sup> century, could anybody tell? How?

# Lecture Thirty—Transcript
# Non-Euclidean Geometry—History and Examples

Welcome back. We are now ready to look at the invention of non-Euclidean geometry. How did it happen? It didn't happen in the way you might expect, where people said: "Let's invent a non-Euclidean geometry" and then worked at it until they succeeded.

Instead, non-Euclidean geometry emerged out of attempts to shore up Euclidean geometry—out of attempts to prove Euclid's Postulate 5 from the others. When we get to the actual moment of invention in today's lecture, it will be a mind-stretching moment—I promise. But first, we have to have a bit of background.

As I've said, questions about Euclid's fifth postulate began in antiquity. But the Greek mathematical tradition faded out in the last days of the Roman Empire. Greek mathematics lived on, though, in the Islamic world.

In the $9^{th}$ century, the Caliph al-Ma'mun established a research institute, the House of Wisdom, in Baghdad. This institute, like the Museum in Alexandria before it, began by collecting all the scientific works around—including Greek scientific works.

These works were then translated into Arabic. Once they had Arabic text, the scholars in Baghdad started a vigorous mathematical tradition. Eventually, mathematics in the Islamic world synthesized the Greek proof-based tradition with the computational traditions of the East—including mathematics from China, India, and Persia, as well as from within the Arabic-speaking world itself.

We still use the Arabic words "algebra" and "algorithm"—a fact that testifies to the influence of mathematics from the medieval Islamic world on modern mathematics. One of the topics mathematicians in the Islamic world addressed was the status of Euclid's fifth postulate.

A number of mathematicians in the Islamic world tried either to prove the fifth postulate, or to replace it by another postulate that would be more self-evident. Here are some of the most important of these people.

I've already mentioned the mathematician and poet Omar Khayyam. Omar used as his first principle that "2 convergent straight lines must intersect."

By "convergent straight lines" he meant lines that got closer and closer to each other. From that principle, Omar was able to prove Postulate 5.

But, of course, Omar—like Lagrange later on—was making an assumption that is logically equivalent to Euclid's postulate, even though he may have liked his own assumption better, or thought it was more self-evident than Euclid's. Later on, Nasir Eddin al-Tusi made a different assumption, also equivalent to Postulate 5. He assumed that a quadrilateral (that's a 4-sided figure) with equal sides whose base angles are right angles must have the top angles also right angles. That was his equivalent assumption.

His son, Sadr al-Din al-Tusi, was also interested in the problem, and Sadr tried to show that Euclid's fifth postulate followed from an assumption similar to the one made by Omar Khayyam about the distance between parallel lines.

For our present purposes, what is most important about Sadr al-Din's work was that it was published in Europe—in Italy in 1594—so his work helped to jump-start the interest of European mathematicians in the problem.

So, from ancient Greece through the 17th century, mathematicians learned more and more about what is logically implied by Postulate 5, and what results are logically equivalent to Postulate 5.

Then, in the 18th century, as we saw in Lecture Twenty-Eight, mathematicians, physicists, and philosophers alike generally thought Euclidean geometry was the absolute model for all intellectual achievement. But, for these Euclid worshippers, there was a problem—because there was this unproved, non-self-evident assumption—Postulate 5—still running around. It's not self-evident, so if it hasn't been proved, there is a flaw in this perfect geometry of Euclid. Something needs to be done.

This 18th-century attitude is exemplified by the title of a book by Gerolamo Saccheri, Italian mathematician. Saccheri thought that he had proved the fifth postulate—he had not, but he thought he had—and so he called his book, *Euclid Freed from All Imperfections*. I'm not going to go into the mathematical details—the books by Katz and by Gray listed in the readings do this very well. But I do want to explain Saccheri's overall strategy because it was novel at its time.

The earlier work, before Saccheri, that I've told you about was all aimed to deduce the fifth postulate—or equivalents to it, like the uniqueness of parallels—directly from the rest of Euclid's postulates. Saccheri, instead, used the weapon of indirect proof.

"Let's assume that Postulate 5 is false," said Saccheri, "and then let's see what follows." A large number of apparently absurd things follow. What Saccheri had done, then—and what later mathematicians who tried the same things, like Lambert and Lagrange—what Saccheri did was to work out a large set of consequences of the denial of Postulate 5, consequences that seemed to Saccheri to be absurd. Let's list a few of these supposedly absurd consequences.

Through an outside point, more than one parallel exists to a given line; or, if you have a quadrilateral where 3 of the angles are right angles, the fourth angle does not need to be a right angle; or, the sum of the angles of a triangle is less than 2 right angles; or, parallel lines are not everywhere equidistant. These conclusions—especially the first and the last ones—seem contrary to our intuition of space, contrary even to our idea of straight lines. But are they "absurd" in the sense required by logic? Are they contradictions, which is what you need for an indirect proof? What counts as "absurd"? Let me tell you a story. In the early 1980s, I spent a year as a visiting professor at UCLA. Then, there were 20,000 students—mostly from southern California, right on top of cultural trends.

At that time in the early 1980s, when people dyed their hair, the colors they used were colors that are actually found in hair. So, a blonde person might dye her hair brown, or a dark-haired person might dye his hair blonde. But one morning, I saw a young man who had dyed his hair bright green. My first reaction was to say: "That's absurd!" But it wasn't really absurd. It was just a violation of what I'd been conditioned to expect. Nowadays, I see students with green hair regularly. So it isn't absurd at all—it's just a different cultural convention.

That little story is supremely relevant to the story of non-Euclidean geometry. Here are all these supposedly absurd consequences of denying Euclid's parallel postulate. So what's the common reaction? They're absurd. They're impossible. Besides, the $18^{th}$ century needs Euclidean geometry. Even the most radical of the $18^{th}$-century people who worked on Postulate 5—Johann Heinrich Lambert—never got completely past the Euclidean mind-set. Lambert was worried that we'd never be able to prove the logical necessity of Euclidean geometry—and he had spent a lot of time trying.

Lambert even briefly suggested that a sphere with an imaginary radius, whatever that might be, that such a sphere could have a geometry where the parallel postulate was false. But Lambert did not conceive of a full-blown non-Euclidean geometry, and neither did anybody else in the $18^{th}$ century.

In the early 19th century, though, something happened. Three people, more or less independently, were struck by the same revelation: "Hey! These conclusions are not contradictory, or illogical, or absurd at all. They are perfectly valid theorems, but in a different geometry—a geometry that is not Euclidean."

That realization was really hard to come to. It required a kind of gestalt switch. Actually, in this course, you and I together have been in the same place when we were proving things about parallels using indirect proof. We just didn't make the switch. When I proved the theorem that "When 2 lines are parallel, the alternate interior angles are equal," I started by assuming that the angles weren't equal.

I wasn't trying to construct an alternate reality, in which lines can be parallel, with the alternate interior angles unequal. But in a sense, that's what I did in that proof. It's Euclid's proof—that's what Euclid did in that proof. But Postulate 5 stopped us from taking that alternate reality seriously.

But just because I couldn't draw the diagrams to make that alternate reality plausible does not mean that the alternate reality doesn't exist. It doesn't mean that there can't be a situation where 2 lines are parallel, and still the alternate interior angles aren't equal. It doesn't mean that there can't be an alternate reality where Postulate 5 is false.

So let's look at the actual inventions of this new kind of geometry—call it an "alternate reality" if you like—and let's look at the people who did it. The 3 inventors were Karl Friedrich Gauss, Janos Bolyai, and Nikolai Ivanovich Lobachevsky, working, respectively, in Göttingen, Germany; in Hungary; and in Russia—all outside the mathematical mainstream centered in France. A brief sociological remark: When people are going to do something really out of left field, it helps when they're not part of the mathematical mainstream—so it's not an accident that they're all kind of on the periphery.

Let's start with Gauss, the first inventor. Gauss was one of the greatest mathematicians in history, and he worked in all parts of mathematics and mathematical physics. Gauss's work on the curvature of surfaces got him interested in the basic assumptions of geometry, and Gauss worked out—and realized that he was working out—a non-Euclidean geometry as early as 1816. Gauss thoroughly understood the logical nature of this new subject. Gauss is the one who named it "non-Euclidean geometry."

But Gauss never wrote up his ideas in full detail, and he never published anything about non-Euclidean geometry. Why didn't he? Gauss was aware that his new geometry was challenging the prevailing Kantian view that space was *a priori* Euclidean. Gauss did not agree with Kant.

Gauss thought that the nature of space was a question for physics, and that the answer should be discovered by experiment. That's a pretty controversial idea for a countryman of Kant.

So one reason Gauss didn't publish his non-Euclidean geometry is that he didn't want to be involved in a controversy. The way he put it to his friends was this: "I didn't want to publish this because I wanted to avoid the wasps buzzing around my ears."

Now on to the next inventor: Janos Bolyai from Hungary. Janos Bolyai's father, Wolfgang Bolyai, was a mathematician himself and a friend of Gauss. The father had been working on trying to prove the parallel postulate, Postulate 5, but he hadn't gotten anywhere, and he found it very frustrating.

So, the father told the son: "I have traversed this bottomless night, which extinguished all light and the joy of my life. I entreat you, my son, leave the science of parallels alone. ... You should detest it more than lewd intercourse." Telling your son stuff like that doesn't always work. So, young Janos continued to work on the theory of parallels. At first he tried it his father's way—and until 1821, Janos, the son, thought that the parallel postulate must be true. But then Janos switched sides. He began to build a geometry that was a series of logical consequences of the denial of the parallel postulate. So there he was—inventing non-Euclidean geometry.

Janos was really excited about what he was doing. He described what he'd done by a biblical metaphor. He said: "I have created a new world out of nothing." Wow.

I wish I could tell you that when young Janos Bolyai published his discovery he became famous overnight—but that is not what happened. He published his discovery as an appendix to a book by his father, Wolfgang Bolyai, in 1831, and quite naturally they sent a copy of the book to Gauss. Gauss wrote back to the father: "I am unable to praise this work, [because] to praise it would be to praise myself."

"What your son has done," said Gauss, "coincides almost entirely with my own thoughts on the subject. I was going to write it all down so it wouldn't die with me. So it's good that I don't have to take the trouble, and I'm glad that the one who did it is 'the son of my old friend.'"

As you can imagine, Janos Bolyai was pretty upset by this. He didn't believe that Gauss had actually done all of this. Janos felt cheated, and he never published anything in mathematics again. It wasn't until the 1860s that all of these matters became public. This was partly because the Bolyais were in Hungary, far from the center of the 19<sup>th</sup>-century mathematical action. It was partly because Janos's work was published as an appendix to another book, and it was partly because Janos didn't push for recognition. Gauss did, in fact, have non-Euclidean geometry—and when Gauss's mathematical papers were published in the 1860s, that became clear.

But Gauss hadn't worked it out as systematically as Bolyai had done. The publication of Gauss's papers included Gauss's letter to Bolyai's father. So after reading this letter, people in the 1860s dug out Bolyai's work and recognized his independent discovery. But he got recognized only after his death.

All the same, Janos Bolyai understood completely how revolutionary what he'd done was and what it might mean. Besides saying: "I've created a new world out of nothing," Bolyai called his new subject "the absolute science of space."

"The absolute science of space" meant geometry without the assumption of the parallel postulate. By "absolute science" Bolyai meant a geometry liberated from the restriction of obeying Euclid's fifth postulate.

Last, but not least, Lobachevsky: Again on the mathematical periphery, Lobachevsky was at the University of Kazan in Russia. One of his professors there was a German, Martin Bartels—who was, by the way, another associate of Gauss. Lobachevsky's non-Euclidean geometry is logically the same as that of Bolyai and that of Gauss. But Lobachevsky did a better job promoting his work. In 1835, Lobachevsky published his own discovery of non-Euclidean geometry, which he called "imaginary geometry." That name, "imaginary geometry," is analogous to the idea of imaginary numbers—those are numbers that involve the square root of $-1$. It's a good analogy because when you study imaginary numbers, what you find is a system that's quite different from the real numbers that we all grew up with—but, nevertheless, it's a logically consistent system, just like non-Euclidean geometry.

Lobachevsky's first publication of his new geometry was in Russian, but his work was soon translated into German and French. So Lobachevsky's was the version of non-Euclidean geometry that got into circulation in the broader mathematical world. The key book was his *Theory of Parallels*, published in

1840. Lobachevsky held the same view that Gauss held—that the question of whether real, physical space was Euclidean or non-Euclidean was an empirical question. Since Lobachevskian geometry—we might as well start calling it that because people often do—since Lobachevskian geometry says that the sum of the angles of a triangle is less than 2 right angles, Lobachevsky said: "We ought to be able to test this if we have a sufficiently large triangle—where, say, the sides of the triangle are the diameter of the Earth's orbit, the distance from one end of the orbit to a star, and the distance of the other end of the orbit to a star."

So Lobachevsky himself—and yet another friend of Gauss, Friedrich Bessel—tried to make such measurements of that huge astronomical triangle, but all that they found was that within the range of measurement error, the sum of the angles of this physical triangle added up to 2 right angles. So, no measurement in the middle of the 19th century was accurate enough to tell whether real, physical space was Euclidean or not.

Nevertheless, Lobachevsky's work demonstrated logically that: Yes, Voltaire, there is more than one geometry. At this point, you ought to be wondering what such a geometry would look like. I mean, fine, it's logically consistent for Postulate 5 to be false. But is there such a thing as a non-Euclidean surface? If so, what would such a surface look like? Let's see. I mean literally—let's "see."

Here is a 3-dimensional computer graph of such a surface—or rather, of a finite part of such a surface because it's an infinite surface. It is saddle-shaped. Here is a physical model of a finite piece of an infinite mathematical surface that has all the Lobachevskian properties; this particular surface is called a "hyperbolic paraboloid." Okay, it's finite, it should go off to infinity like this, and then off to infinity going down on both sides.

But wait! This surface? What is a straight line on this surface? Let's borrow an idea from Archimedes, which he made a postulate, but which I will make a definition. On this physical model, we will define a straight line as "the shortest distance between 2 points on the surface." I mean, that's the same definition we'd use on a Euclidean surface. Let me show you on the table. All right, the shortest distance between my 2 thumbs. Stretch string on the table—that's the shortest distance on this plane surface. So, now I'll stretch the string to make some straight lines—that is, shortest distances—on this physical model. All right, like here, between this point and this point. I've marked a bunch of these permanently with these stretched threads. We will use Euclid's definition of parallel lines as 2 "shortest distances"—2 lines—on that same surface that never meet.

Now we're ready for a truly exciting visual demonstration. Look at this one and this one. You can have 2 lines through a single point parallel to a third line. See? As you can also see, parallel lines on this surface—like this one and this one—are not everywhere equidistant. All that happens because Euclid's Postulate 5 is false on this surface. You can see that Postulate 5 is false on this surface. All right, this angle here is a right angle. This red angle here (let's get this thread in the right place) is less than a right angle, so these 2 angles add up to less than 2 right angles—and yet these lines don't meet. See, this converges towards this line for a while, and then it starts to diverge again. They don't meet. Postulate 5 is false on this surface.

If you want to test some of this, you can buy yourself a whole can of non-Euclidean surfaces and do these experiments yourself. You take a manufactured potato chip—be very gentle with it; they break easily—and stretch little pieces of thread the way I've done here to determine the shortest distance between 2 points. Then, you'll be able to see for yourself that parallel lines—that is, shortest distances that don't meet—parallel lines on such a surface are not everywhere equidistant, and also that there can be more than one parallel to a given line through an outside point. Okay, you see the 2 parallels to this line that go through this point, and you can certainly see that these parallel lines aren't everywhere equidistant.

So, Euclid's logic seems to be okay so far—that is, theorems whose proofs required Postulate 5 aren't always true on this surface. Let's now test some of the theorems that we have actually proved in this course. Theorems 27 and 28 didn't need Postulate 5 in their proofs, so they should still be true; Theorems 29 and 30 might well be false. So, let's look.

First, 27 and 28—the ones that should be true even on a surface where Postulate 5 is false. Proposition 27 says: "If the alternate interior angles are equal, then the lines are parallel"—then the lines don't meet. So, let's have a look. This black angle and this black angle, these guys—they're equal. All right? Two equal angles. They happen to be right angles, but this angle is equal to that angle. They're alternate interior angles, so these lines don't meet—so they are parallel. Proposition 28 says: "If the corresponding angles are equal, then the lines are parallel"—right; that happens here, too. This and this are corresponding equal angles. This line and that line do not meet.

We're good so far. What about those theorems whose proofs did need Postulate 5? What about Proposition 29—if 2 lines are parallel, must the corresponding angles be equal? Let's have a look. This is the shortest distance. This is the shortest distance. Here are corresponding angles—this guy here and that red one there; they're not equal. They're not equal. That part of Proposition 29 is false. Likewise, if the lines are parallel, must the alternate interior angles be equal? Look here, this one and this one. Alternate interior angles—they're not equal; it's false.

How about Proposition 30: "Two lines parallel to a third line are parallel to one another"? These guys are both parallel to this one, but they're certainly not parallel to one another. They meet. Looking at this little triangle that I've drawn—I did a shortest distance across here. I think you can see that the angles look smaller than what we'd have with a flat triangle with the same corners—and so, the sum of the angles of that triangle add up to less than 2 right angles.

So Euclid got it completely right. He never imagined a non-Euclidean space—but his logic works for it anyway. The theorems that are supposed to be true without Postulate 5 are true on this non-Euclidean surface; the ones that aren't necessarily true without Postulate 5 aren't true on this surface. Euclid has got it right.

Non-Euclidean geometry is out there in the real world, on a much larger scale than potato chips. This surface has been of interest to modern architects—and not just in famous things in big cities. Here's a gas station from 1958 in Kansas, for instance—that's a hyperbolic paraboloid.

In Lecture Thirty-Three, I'll talk a lot about the impact of these new geometric ideas on modern art. But for right now, I like this example because it isn't a major architectural landmark. It's just an ordinary gas station in Wichita, Kansas, and so it shows how this once-revolutionary idea has become assimilated into the modern world. At this point, you may have an objection. You might say: "Look here, Professor Grabiner, this so-called non-Euclidean surface isn't really a plane like Euclid's. That's a 3-dimensional object." This is why it was so hard to invent non-Euclidean geometry. We need to rethink a lot of our ordinary ideas—for instance, redefining a straight line as the shortest distance between 2 points on a particular surface. But Euclid's definition of a straight line—"A straight line is one that lies evenly with the points on itself"—that wasn't one of his best definitions, so maybe you were willing to let me get away with that one.

But let's look at the surface for a minute to see that it's a surface and not a 3-dimensional object. What we mean by a surface is that, if you're at a point on the surface, you can go left, you can go right, you can go forward, and you can go backwards—but you can't go upward from the surface, and you can't tunnel down through it.

When I said that the shortest distance between 2 points on this surface could be determined by this stretched thread, you didn't say for this one: "Why didn't you fly from here to here?" or, for this one, from here to here, you didn't say: "Why don't you just tunnel through the surface? No, it's a surface. Your shortest path—stay on the surface, and it's a surface whose geometry is non-Euclidean.

There is another possible objection, this one from logic. I've kept saying that "If p, then q" does not imply "If not p, then not q." So you might say: "Well, just because the fifth postulate is false on the surface, that doesn't mean that the propositions Euclid proves using the fifth postulate must be false, does it?" Logically speaking, "No." In fact, though, those propositions are—as we've seen on this model—those propositions are not always true on this surface. Actually proving what is true and what is not true in Lobachevskian geometry is an important part of the achievement of Gauss, Bolyai, and Lobachevsky.

I've been acting as though all they did was to realize that the denial of Postulate 5 led to things that were not absurd, but valid theorems in a non-Euclidean geometry. But Gauss, Bolyai, and Lobachevsky did much more. They had to do, for non-Euclidean geometry, what Euclid had done for Euclidean geometry—and they did. They proved the theorems that show what a non-Euclidean world was like.

All of this, I hope, has been pretty exciting. But what I've done so far raises a large number of questions. Here are 3 of the most urgent questions. First: We've looked at a geometry that denies Postulate 5. Could there be yet another geometry—a geometry that denies one of the other postulates of Euclid? If so, what would that be like? Second: I've shown you that there's a 2-dimensional non-Euclidean geometry—a geometry on a surface. But we live in a 3-dimensional world. So could there be a 3-dimensional non-Euclidean geometry? If so, how could we visualize it? How could we visualize a 3-dimensional non-Euclidean geometry? What would Kant and his followers say about that? Last question: Does non-Euclidean geometry have any practical value, or is it just an intellectual curiosity? I will address all of these questions in the next lecture.

# Lecture Thirty-One
# Non-Euclidean Geometries and Relativity

**Scope:** As we have just seen, non-Euclidean geometry of the Lobachevskian type describes negatively curved surfaces. Now we consider positively curved surfaces as well. Building on the work of Bernhard Riemann, we discuss the many ways a space can be characterized. We exhibit a model of a non-Euclidean surface of the Riemannian type. We now distinguish between 3 types of surfaces—Euclidean and flat, Lobachevskian and negatively curved, and Riemannian and positively curved—and we look at the sum of the angles of a triangle, and the number of parallels to a given line through an outside point, on each of these. We conclude with the remarkable story of how Albert Einstein used Riemannian geometry in the general theory of relativity.

# Outline

I.  It seems now that any surface can have a geometry in which we define a straight line as the shortest distance between 2 points on that surface.

  A.  The shortest distance between 2 points on the globe is the shorter arc of the great circle connecting those points.
    1.  A great circle is a circle on the globe whose diameter is the same as the diameter of the globe.
    2.  The equator and the meridians of longitude are great circles.
    3.  Parallels of latitude also are circles on the globe, but smaller.

  B.  We give an informal proof that the arc of the great circle is in fact the shortest path between 2 points on the globe.

II.  The geometry of a sphere is usually called Riemannian geometry, after the German mathematician Bernhard Riemann.

  A.  Riemann generalized the notion of "surface" and the notion of "distance."

  B.  By doing this, Riemann brought non-Euclidean geometries into the mainstream of mathematics.

**III.** The geometry on the surface of the sphere was well understood from antiquity. Why, then, did past mathematicians not realize that the surface of a sphere is non-Euclidean?

   **A.** First possible answer: The sphere is not non-Euclidean; it is a Euclidean object in 3-dimensional space.

   **B.** Second possible answer: We can refuse to recognize that the sphere has anything to do with the debate about the nature of space or the status of Postulate 5.

   **C.** Third possible answer: We can refuse to think that the laws for the surface of a sphere are a geometry. This attitude prevailed until the 19th century.

**IV.** We return to the 3 different kinds of geometries we have discussed, and we compare them mathematically.

   **A.** One way to compare them is to look at the 3 different kinds of surfaces involved in the geometries of Riemann, Lobachevsky, and Euclid.

      **1.** The sphere is said to be positively curved.

      **2.** The Lobachevskian surface is said to be negatively curved.

      **3.** The Euclidean surface has zero curvature (flat).

   **B.** Another way to compare these 3 types of geometry is to see what happens to the property of parallels highlighted by Playfair's axiom.

      **1.** In Riemannian geometry, given a line and an outside point, you can draw no parallels to the given line through the outside point.

      **2.** In Lobachevskian geometry, you can draw more than one parallel.

      **3.** In Euclidean geometry, you can draw exactly one parallel.

   **C.** Yet another way to compare these 3 is to recall which of Euclid's postulates are not valid on a particular surface.

      **1.** On the sphere, Postulate 2 does not hold.

      **2.** On a Lobachevskian surface, Postulate 5 does not hold.

      **3.** In Euclidean geometry, all of the postulates are valid.

   **D.** In our last comparison, we compare the sums of the angles of a triangle in each of the geometries.

      **1.** In Riemannian geometry, the sum is greater than 2 right angles.

**2.** In Lobachevskian geometry, the sum is less than 2 right angles.

**3.** In Euclidean geometry, the sum is equal to 2 right angles.

   **E.** The important point is that these geometries are mutually contradictory.

**V.** We might say that each of these geometries is equally certain. But which is true of the real world? This was the question being asked in the 19th century.

   **A.** Until the early 20th century, Euclidean space was seen as corresponding to the real world, while non-Euclidean geometries, though internally consistent, were thought of as fictions.

   **B.** This changed when Albert Einstein started working on his theories of relativity.

**VI.** We briefly explain, following Einstein's own popularization, the basic ideas of the general theory of relativity, especially with respect to gravitation.

   **A.** The key idea of the special theory of relativity is that no physical experiment can be performed to tell whether we are accelerating at a constant rate in the absence of any forces or at rest in a gravitational field.

   **B.** We draw from this the consequence that, in the presence of a gravitational field, the path of a ray of light is not straight but curved.

   **C.** The curvature of space is altered by the presence of a body with mass, which causes the bending of a light beam.

   **D.** So, the light rays from stars are bent as they pass close to the Sun.

   **E.** We describe the successful experiment set up to measure this effect.

**Essential Reading:**

Einstein and Infeld, *The Evolution of Physics*, 220–260.

**Suggested Reading:**

Gray, *Ideas of Space*, 210–225.

**Questions to Consider:**

1. The geometry of the sphere has been known for a long time, and the spherical trigonometry that describes the Earth was well understood by Ptolemy of Alexandria (100–178 C.E.) and developed fully in the Islamic world, in part to find the direction of prayer to Mecca from any part of the globe. Why, then, did nobody seem to care that geometry of the surface of the spherical Earth did not obey Euclid's postulates?

2. The general possibility that there could be other geometries, with other sets of postulates, seems obvious now; why did it take so long for people to get the idea?

# Lecture Thirty-One—Transcript
## Non-Euclidean Geometries and Relativity

Hello again. Back in Lecture Twenty-Five, we started this part of the course in Voltaire's world, where there was just one geometry: Euclidean geometry. But in Lecture Thirty, we saw that on a different kind of surface, there was a different geometry. Now, we should ask: "What if we define a straight line as the shortest distance between 2 points on any surface. Then does every different kind of surface have its own geometry?" It turns out that the answer is "Yes." But before we do any more generalizing, let's look at what will be a surprisingly familiar example: a sphere—not just any sphere, either: our sphere, the globe.

Let's look at the geometry on the globe, treating it as though it were a perfect sphere. First, we need to define a straight line. It's supposed to be the shortest distance between 2 points. So let's stretch a string on the globe and see, for instance, what's the shortest distance between Vancouver and London? The shortest distance between 2 points on a globe is the shorter arc of the great circle connecting those points—a great circle is a circle that has the same diameter as the globe. Like the meridians of longitude, or like the equator, those are great circles—not parallels of latitude; the parallels of latitude are circles, right, but they are smaller, and they aren't shortest distances. That's why you fly from Vancouver to London on a path that curves towards the pole—that's a great circle—and not along their shared parallel of latitude—that's longer.

There's a neat way to see why the shortest distance between 2 points on the globe has to be a great circle, rather than any of the other circles you can draw on the globe. Imagine that we slice into the globe with a plane. The intersection of the globe and the plane that slices it will always be a circle. Let me illustrate this with an orange and a knife. An orange is more or less a sphere, and a knife is more or less a plane. So, we slice through the orange, and wherever we slice through the orange, we get a circle. This is a small circle. Here I slice through it, we get something whose diameter is the same as the diameter of the orange. That's a great circle.

If we have 2 points on the globe and lots of circles passing through those 2 points, which of all those circles gives the shortest distance? It's the arc of the circle that has the greatest radius, the arc with the greatest radius—as this picture illustrates.

Of all the circles on the globe, the circle that has the greatest radius, or the greatest diameter, is the one whose diameter is the same as the diameter of the globe itself. So, this shortest distance between 2 points on a globe is on the arc of that greatest circle.

Let's look at the set of great circles on the sphere. One thing should strike us immediately if you look at, say, the meridians of longitude, which are great circles, and the equator: Every pair of great circles on the sphere intersects.

Wait! So there are no parallel lines at all? Yes, we have parallels of latitude, but remember the latitude lines are not the shortest distances—they aren't our straight lines. The great circles, which do play the role of shortest distances or straight lines on the globe, the great circles all intersect.

But wait a minute—didn't Euclid prove the existence of parallel lines back in Book 1, Proposition 31? Remember in that theorem, Euclid proved that we can construct parallel lines? We ought to have confidence in Euclid's logic by this time—so, if the theorem that we can construct parallel lines is false on the sphere, since there aren't any parallel lines, one of the earlier results we used to prove that theorem must also be false. How was it that we proved that we could construct a parallel? We used the theorem that says that: "If the alternate interior angles are equal, then the lines are parallel"— Proposition 27. Is that true on the globe?

Let's look at what happens when 2 great circles, each one a meridian of longitude, intersect the equator, which is another great circle. So, here's the equator. Here's a meridian of longitude. Here's another meridian of longitude. Okay, the meridians of longitude are perpendicular to the equator because they form equal angles with the equator on either side; those are right angles.

So here are 2 lines that cut a third line such that the alternate interior angles are equal. The meridian of longitude makes a right angle here. The other meridian of longitude makes a right angle down there. But the lines aren't parallel because the 2 great circles intersect.

All right, so that theorem: "If the alternate interior angles are equal, the lines are parallel"—that's false on the sphere. What did we need to get that theorem? You'll recall that was an indirect proof, and we used in that proof the theorem that: "The exterior angle of a triangle is greater than the opposite interior angle." That theorem also is false on the globe. The same picture shows us that. The exterior angle is a right angle, and the exterior angle equals the opposite interior angle because they're both right angles. So that theorem—Proposition 16—is false on the sphere.

So, what did we assume in proving that theorem that: "The exterior angle is greater than the opposite interior angle"? What did we assume in proving that theorem that might not hold on the sphere? Remember Euclid's diagram for the proof of that theorem. In that proof, we took the line BE, and we extended it to double its length—to BF, so that BE was equal to EF. In order to do that, we needed Postulate 2, which says that: "A line [in particular, BE] can be extended to any desired length [in particular, double its length]." Is Postulate 2 true on a sphere?

The answer is "No." The great circle is finite. You can start out extending the line—say, to the equator—but there's a limit to how far you can go. In particular, if you try to extend the line from the North Pole to the South Pole to double its length, you're back where you started! So Postulate 2 is false on the sphere—and, therefore, the propositions that depended on Postulate 2 are not necessarily true on the sphere. We can see that they, in fact, are not true on the sphere. In particular, there are no parallels—all the great circles intersect, like the meridians of longitude and the equator.

So, now we have still another geometry on a different surface. This new geometry starts by denying the assumption that a straight line can be made as long as you want. In this geometry, we see clearly that the sum of the angles of a triangle on the sphere is greater than 180 degrees. Look again at this picture. The meridians of longitude are perpendicular to the equator— this is a right angle, and this is a right angle, so there are 2 right angles already, plus whatever is at the top. The sum of the angles of a triangle on the sphere is greater than 2 right angles.

The geometry like that on the sphere also has a name. It's called "Riemannian geometry," and it's after a person, the German mathematician Georg Bernhard Riemann. But Riemann did much more than just imagine one more geometry. Riemann generalized the notion of "surface" beyond just the examples that we've seen and also generalized the notion of "distance." He allowed a space, called a "manifold," to have any number of dimensions. By doing all of that, Riemann brought non-Euclidean geometries—that's plural, non-Euclidean geometries—into the mainstream of mathematics. The geometry on the surface of the sphere that's named after Riemann is just one example.

But wait a minute. The geometry on the surface of the sphere was well-understood by the Greeks—I mean, the earth is a sphere after all; they made maps of the world. The geometry in the sphere was even better understood by the Muslims, who could find the great-circle direction—that is, the shortest distance between 2 points on the globe, where one point is where they were and the other was Mecca, because that's the direction of Islamic prayer, along the great circle connecting where you are to Mecca. They could find that from any point on the globe.

So, you may be wondering, why didn't the Greeks and why didn't the Muslims all realize that what they had here was a 2-dimensional non-Euclidean geometry? That's a really good question. Of course, they realized that the geometry of the sphere isn't the geometry of the plane. So why didn't the Greeks or the Muslims say: "Aha! A non-Euclidean surface"?

It's always hard to answer questions about why somebody didn't discover something. It's impossible to know for sure. But for this question—why didn't past mathematicians realize that the surface of our world is a non-Euclidean surface and, therefore, there can be non-Euclidean geometries?—there are several possible answers.

Here's the first possible answer: The sphere isn't non-Euclidean. It's a Euclidean object in a 3-dimensional Euclidean space.

Second possible answer: We can refuse to recognize that the sphere has anything to do with the debate about the nature of space or the status of the only postulate that people were worried about, which is Postulate 5. Remember Lagrange, who for a while thought he'd proved Euclid's parallel postulate?

Lagrange, himself, conceded that you didn't need Euclid's parallel postulate to study triangles on a sphere. But what was at issue for all these people who tried to prove Postulate 5 from the others was not whether you can imagine alternate sets of rules. What was at issue was the nature of real space. By the 18th century, that question was always answered in favor of Euclid. You'll remember that Lagrange, himself, needed the Euclidean properties of space for his major achievements in physics. The fact that the surface of a sphere is curved doesn't change the 18th-century belief that real, 3-dimensional space must be Euclidean. Besides, on one level, Lagrange is right. Postulate 5 isn't false on a sphere—it's just entirely unnecessary. You don't need any conditions for 2 lines to meet. They all meet. It's certainly true that [if] the sum of the interior angles on the same side are less than 2 right angles, the lines meet—because all the lines meet.

Third possible answer: We can just refuse to think that the laws for the surface of the sphere are a "geometry." After all, that's a 19th-century concept—that's Riemann's idea, that every surface has its very own geometry and its very own distance function. That more general point of view wasn't even a live possibility until the time of Gauss, Bolyai, Lobachevsky, and Riemann.

But after Riemann, it's a whole new ballgame. To see the new game, let's return to the 3 types of geometry that I've discussed, and let's compare them mathematically in kind of a systematic way.

One way to compare them is to look at the 3 different kinds of surfaces (we're just doing surface geometry) involved in the geometries of Riemann, Lobachevsky, and Euclid. The sphere is said to be positively curved, the Lobachevskian surface, or the potato chip, or the saddle. The Lobachevskian surface is said to be negatively curved—that's because it has 2 opposing directions of curvature—this way and this way. The Euclidean surface has zero curvature—it's flat.

Another way to compare these 3 types of geometry is to see what happens to the property of parallels highlighted by Playfair's axiom. In Riemannian geometry, given a line and an outside point, you can draw no parallels to the given line through the outside point. In Lobachevskian geometry, like on the saddle surface, you can draw more than one parallel. In Euclidean geometry, the flat plane, you can draw exactly one parallel.

Another way, still, to compare these 3 is to recall which of Euclid's postulates we do not have on a particular surface. On the sphere, it's Postulate 2 that doesn't hold—that's the one that says you can extend the line as long as you want. On a Lobachevskian surface, it's Postulate 5—the postulate about parallels—that doesn't hold. In Euclidean geometry, we have all 5 postulates.

For the last comparison I want to make today, let's add up the angles of a triangle. In Riemannian geometry, the sphere—the sum of the angles of a triangle is greater than 2 right angles. In Lobachevskian geometry—the sum of the angles of a triangle is less than 2 right angles. Of course, in Euclidean geometry, as we've seen Euclid prove and Kant prove—the sum of the angles of a triangle is 180 degrees: 2 right angles.

I could go on doing this kind of comparison. For instance, Riemannian geometry is sometimes called "elliptic geometry," and Lobachevskian geometry is sometimes called "hyperbolic geometry." But the main point of doing this comparison is not for you to memorize names and properties. It's to emphasize a most revolutionary point: These 3 geometries are mutually contradictory.

Lots of famous philosophers and scientists—Plato, Aristotle, Descartes, Newton, and so on—lots of these people have said geometry is certain because it proves everything logically from self-evident assumptions. But now we've got 3 mutually contradictory geometries: with one parallel, with more than one parallel, with no parallels.

One of the other things we've learned from Riemann is that we can have as many dimensions as we want. So, for instance, there can be 3-dimensional geometries (plural) of 3-dimensional spaces that are positively curved; the space is positively curved or the space is negatively curved; or the space has zero curvature. We have 3-dimensional geometries (plural).

You might say each of these geometries is equally certain. After all, the logic is right in all of them. But because they have different basic assumptions, these geometries contradict each other. So at most, one of them can be true of the real world. Which geometry is true of the real, 3-dimensional world we live in? Maybe none of them is true of the real world.

So, that's where we are in the 19th century. There's a lot of interesting philosophy that addresses those questions, and we will look at that philosophy together in the next lecture. But now, let's see what happened historically to the question: Which geometry describes the real world?

You'd think the answer was easy. We live in Euclidean space so far as we can tell, right? So, for 3 dimensions, aren't these geometries just intellectual curiosities? They're consistent all right, but they aren't real. In a way, it's like fiction—like the way William Shakespeare constructed a consistent world in his play, *Hamlet*. Once I was on sabbatical in Denmark, and I took the occasion to go to the castle—Kronborg Castle—in the town of Helsingor. This castle is where the action of the play, *Hamlet*, is set. But when I was there, I did not expect to see the ghost of Hamlet's father up on the battlements. I did not expect to meet Hamlet and Ophelia inside the castle. My expectations were right: I didn't. So, in the same way, non-Euclidean geometries may be internally consistent, but we really shouldn't expect them to correspond to reality, should we? They're fictions. That was more or less how things were until the 20th century—until Albert Einstein started working on his General Theory of Relativity. Let me very briefly tell you what's involved here, and then we'll see—amazingly—what non-

Euclidean geometry has to do with reality. First, I want to talk about special relativity—and then, general relativity.

The key idea of Einstein's Special Theory of Relativity is that there is no experiment that you can do that will tell you whether you are at rest, or moving in a constant speed in a straight line. With respect to the physics of motion, that was known already to Galileo. All right, I'm not moving, and I can drop a piece of chalk, and it falls right in my hand. Suppose I'm in an airplane going 600 miles an hour, I can drop a piece of chalk—and it still lands right in my hand. In fact, I'm on a moving earth (so are you) that is traveling through space at 18 ½ miles per second, and in the quarter-second that it takes me to drop this piece of chalk to my hand—the earth goes like 4 or 5 further miles. I drop the piece of chalk, and it still falls right in my hand. It doesn't go slamming through the wall being left behind by this motion at 18½ miles a second.

That was one of the arguments that Galileo used to show that the earth can move and the physics of motion would still be the same. Einstein's Theory of Special Relativity picks up on this and shows that the laws of electricity and the laws of magnetism, and the measurements you make of the speed of light, are also always the same—whether we're moving at a constant speed in a straight line, or whether we're at rest. That's special relativity.

Now for Einstein's General Theory of Relativity. The key idea of the General Theory of Relativity is that there is no experiment that you can do that will tell you whether you are at rest in a gravitational field, or moving somewhere where there is no gravitational field, but with a constant acceleration. Let me explain why somebody might come up with a theory like that. The force of gravity on an object—say, a piece of chalk—depends on the mass of the object—say, the mass of the chalk. That's part of Newton's law of gravity. But the force required to put the chalk into motion in the first place—that is, the force required to accelerate it—also depends on the mass of the chalk.

This is something you can easily test for yourself. You've got 2 books here. Which one is heavier—which one has the greater mass? I don't have a scale, so what I do is I heft them. I can feel in my arm that this one needs a much greater force to be put into motion. So, I can tell that this one has a greater mass. That's Newton's second law: Force = mass times acceleration. The more force to produce the same acceleration, the more mass it has. These 2 things exactly balance each other out in the law of gravity.

Something has twice as much mass? That makes it twice as hard to put into motion. But because it has twice as much mass, the force of gravity on it is also twice as large. So, the result is that bodies, 2 bodies, with different masses are accelerated at exactly the same rate—if you neglect air resistance. So, this heavy rock and this lighter piece of chalk, if I drop them both at the same time, will hit the floor at the same time.

Einstein reasoned like this: If the acceleration due to gravity is the same no matter what the mass of a body is, then gravity is really a geometrical property. A geometrical property! Einstein worked out what the equations of this geometrical property had to be, and he asked a mathematician friend whether there were any mathematical structures that had exactly the properties that he wanted. "Why yes," his mathematician friend said (listen to this): "a non-Euclidean geometry of the Riemannian type."

So, here is what Einstein thinks is going on in gravity. An object with a large mass, like the sun, makes the space around it curved. That curvature is what makes objects deviate in their path from Euclidean straight lines— like, why the planets orbit the sun. They follow the shortest paths in this curved space—which are not Euclidean straight lines, but curves.

That's a pretty amazing idea. Let's see how Einstein himself explained general relativity to a non-mathematical audience. Imagine that you are in an elevator out in space, with no massy bodies anywhere around, and that the elevator is being accelerated upward at an acceleration of 32 feet per second per second—that's the same acceleration that falling bodies experience in the earth's gravity. So, the elevator is being accelerated upward.

Remember what Newton said about real accelerations producing real forces? So, there's no gravitational field around, but if you were in that upwardly accelerating elevator, you would feel yourself pushed down towards the floor with a force of what is called "one G"—one gravitational force.

If you are in this elevator accelerating upward, if you let a ball go at the end of your hand, the floor will accelerate upwards to meet it—so with respect to you, it looks as though it's falling. It will look exactly like this. I think I'll use the table because the table is attached to the floor. So, here we go. It would look exactly like this.

Did you see the table accelerating upward to meet the ball? If you toss a ball up into the air, again the elevator is accelerating upward. The ball only has the initial speed you gave it—so the elevator, the floor, the table, are going to overtake it. So, inside the accelerating elevator, the ball looks as though it goes up, slows, and stops, and then starts to fall down because everything else is accelerating up to catch it. It would look exactly like that. In other words, you see and feel exactly the same things you would see and feel if, instead of being in an elevator accelerating upward, you were in a room that's at rest inside a gravitational field—like the room I'm in and probably the room you're in.

But wait. Here's Einstein. Maybe I can devise an experiment that will tell the difference between the accelerating elevator, and the room in the gravitational field—okay, accelerating elevator with no gravitational force around, and the room in the gravitational field. What I'll do is I will shine a light beam through the side of the elevator, like this. If the elevator is at rest, like in the left-hand picture, the light beam comes in and goes straight through. If the elevator is accelerating, though, the path of the light beam will be curved—as in the picture on the right—because it comes in there and the walls are accelerating upward while the light beam is going across the elevator. But you'd think, if we were in a room at rest in a gravitational field, the light beam should just go straight through. Right? "No," says Einstein.

I know, I know, all the way back to the Greeks, everybody says light travels in straight lines—Euclidean straight lines. That's why objects have sharp shadows. That's how come we can have Renaissance art with geometric perspective—light goes in straight lines. But Einstein does not agree. Instead, light—according to Einstein—always travels in the shortest path. In Euclidean space, space with zero curvature, that shortest path is a Euclidean straight line. But if the space is curved, the shortest path—the path of the light beam—will also be curved. According to Einstein, as I said earlier, the mass of the earth or the mass of the sun causes space to curve—and, therefore, the light beam going by one of these massy objects will follow a curved path. So, the light beam in the elevator at rest in the gravitational field will follow a curved path, just the way the light beam in the accelerating elevator did—so, Einstein's general relativity.

But given how fast light travels—light travels 186,000 miles per second—this curving of the path is a very small effect. How could we possibly measure it? Einstein found a way. The sun is a very massy object, indeed, and so space around it has got to curve a fair amount. So, suppose there is a star that, from our point of view, is just a little bit to one side of the sun. The sun changes its position with respect to the stars as the earth goes around the sun. So, if we could compare the position of that star relative to another star—that is, compare the angle between the 2 stars as seen from the earth, once when the sun was between us and the 2 stars, and once when the sun was not between us and the 2 stars—the angle we measure would be different. This picture, much exaggerated, shows the basic idea. If the sun were elsewhere, its mass wouldn't bend the light beams. Look at where the stars are, where they actually are, and what the angle between them ought to be according to Euclid's geometry and Newton's physics, if the light goes in a straight line to us. But if the light beams were to curve, the angle we would perceive between the stars would be larger. So, if we could measure the angle between 2 stars when we see them both at night, and then measure the angle again when they're just on either side of the sun, we'd measure a different angle—that is, if Einstein is right, we'd measure a different angle.

That's a great idea, Professor Einstein, but there's a problem: The sun is too bright. When the sun appears right next to the stars, you can't see the stars at all. But you can see a star right near the sun under very special circumstances—when there's an eclipse of the sun. So, in 1919, when an eclipse was predicted for the South Pacific, an international expedition of scientists went down there to test Einstein's prediction.

Here's what Einstein predicted that they would see in 1919. Einstein was right. Within the limits of accuracy of measurement, the light was bent just the way Einstein's General Theory of Relativity predicted. So it turns out, physicists now think, that space around massive bodies is curved—positively curved. Riemannian geometry is needed for general relativity. Einstein's explanation is that the mass of a star, or sun, or planet distorts the space around it—curving it. We detect the shape of space by the effect it has on rays of light. The light's path is the shortest possible in that curved space. Mathematicians first imagined that there could be intellectually a non-Euclidean world. Turns out that we live in one. That's pretty remarkable.

Let me close this lecture with 2 historical questions raised by these amazing developments. First: How did philosophers respond to these new non-Euclidean spaces? Second: How did artists respond to non-Euclidean space and to relativity? These questions will provide the topics for our next 2 lectures.

# Lecture Thirty-Two
## Non-Euclidean Geometry and Philosophy

**Scope:** We saw in earlier lectures that philosophers and scientists, from Voltaire to Pierre-Simon Laplace and Immanuel Kant, seemed to require space to be Euclidean. We will now make precise the way in which non-Euclidean geometry challenges those views. We will see how some thinkers regarded non-Euclidean geometries as mere intellectual curiosities and held that real space was nonetheless Euclidean. Hermann von Helmholtz, though, provided a different view of geometry, arguing that we can learn to order our perceptions in a non-Euclidean, 3-dimensional space. We will defend the plausibility of Helmholtz's view and then conclude with some psychological experiments and cross-cultural research about the many different ways that people have thought about space.

## Outline

I. For 2000 years, philosophers were committed to the idea that space had to be Euclidean.

   A. The Euclidean nature of space was linked to a wide range of important properties of the natural and philosophical worlds.
      1. First, there was the nature of the straight line.
      2. Then, there was Gottfried Wilhelm Leibniz's principle of sufficient reason.
      3. Also, there was Isaac Newton's idea of absolute space.
      4. As we have seen, people of the stature of Voltaire and Pierre-Simon Laplace had strong arguments using the Euclideanness of space.

   B. The most important set of philosophical ideas on behalf of Euclidean space, though, were those set forth by Immanuel Kant. He said that space was the unique form of all possible perceptions, that we order our perceptions in our pure a priori intuition of space—and that space was Euclidean.

   C. If non-Euclidean geometry is to become philosophically respectable, Kant's arguments must be answered.

**II.** These questions were addressed by Hermann von Helmholtz, who found them compelling due to his interest in psychology, physics, and the physiology of vision.

    **A.** In his essay "On the Origin and Significance of Geometrical Axioms," he asked what geometry 2-dimensional beings living on the surface of a sphere would invent.

    **B.** For such beings, imagining moving out of their surface into a third dimension would be like a blind person's imagining color.

    **C.** Helmholtz demonstrates visually that we can learn to order our perceptions in a 3-dimensional, non-Euclidean space.

        **1.** Look at the reflections in a convex mirror in Escher's print.

        **2.** Notice the non-equidistance of the parallel top and bottom of the bookcase.

    **D.** Helmholtz asks, "How can we tell which of these possible worlds is the real one?" He constructs a relativistic argument.

**III.** Another perspective on these questions comes from France, from the great French mathematician Henri Poincaré.

    **A.** Poincaré disagreed with both Kant and Helmholtz.

    **B.** Poincaré said that "geometrical axioms are conventions."

**IV.** In the English-speaking world, a seminal thinker about geometry was the mathematician and philosopher W. K. Clifford.

    **A.** Clifford argued that Lobachevsky's work revealed that Euclid comprises only a small part of geometry.

    **B.** He pointed out that space can only be proven to be statistically flat, and only for "as far as we can explore."

    **C.** Clifford used non-Euclidean geometry to reach the same conclusions about the limits of knowledge that had been reached by statisticians and probability theorists.

    **D.** Some thinkers still stood by Euclid, but Clifford's approach was more compatible with the empirical nature of 19th-century science.

**V.** Because of non-Euclidean geometries and noncommutative algebras, the philosophy of mathematics has undergone a major change since the time of Plato.

    **A.** Plato held that the world of sense experience was an imperfect model of the realm of the forms.

**B.** Aristotle thought that the objects of mathematics were idealized abstractions from the world of sense experience.

**C.** Now it appears that there are many possible mathematical models for reality, of which at most one can be correct, and that some of these mathematical structures are invented with no reference to the world of sense experience.

**D.** We look at the views on this of the first great American mathematician, Benjamin Peirce.

**VI.** Helmholtz's psychological investigation of whether we can order our perceptions in any other way than the 3-dimensional space of Euclid suggests a range of other important questions.

    **A.** Cultures other than those whose idea of space was formed by Euclidean geometry talk about space in ways quite different from that of Newton or Kant.

        **1.** The use of fixed coordinate systems appears in other cultures.

        **2.** The Leibnizian idea of space as just the relations between objects appears in other cultures as well (and is used by GPS systems when they give directions).

    **B.** Questions about geometry are deeply involved in major modern discussions, not only about philosophy but also about psychology and about the nonuniversality of our cultural picture of the world.

**Essential Reading:**

Von Helmholtz, "On the Origin and Significance of Geometrical Axioms."

**Suggested Reading:**

Levinson, "Frames of Reference and Molyneux's Question."

Poincaré, *Science and Hypothesis*.

Richards, "The Geometrical Tradition."

Wagner, *The Geometries of Visual Space*.

**Questions to Consider:**

**1.** What might Voltaire say about non-Euclidean geometries and about the many different cultural views of space?

**2.** If mathematics is just "the science that draws necessary conclusions," why is it any different from, say, chess, which also is just about working out the consequences of a predetermined, arbitrary set of rules?

# Lecture Thirty-Two—Transcript
# Non-Euclidean Geometry and Philosophy

So, all of a sudden we're living in a non-Euclidean space. What are philosophers going to say about that? For 2000 years, Euclid's was the only geometry. Philosophers respected it not just for its logic—although geometry was certainly respected for that—respected it not just for its useful applications to everything from architecture to warfare, although it was respected for that, too. Philosophers respected Euclidean geometry for giving a self-evidently true account of the world—self-evident and true premises, good logic, conclusions both true and useful, the highest achievement of human reason.

Philosophers, as you'll recall, had linked the Euclidean nature of space to a wide range of other important properties of the world of nature and the world of thought. There was the nature of the straight line—and, according to Newton's first law, the fact that motion in the absence of forces followed straight lines. There was Leibniz's principle of sufficient reason. There was Newton's idea of absolute space. There were all the philosophers, especially in the 18th century, who used Euclidean geometry and the Euclidean nature of space as a cornerstone of their systems of thought—from Voltaire to Laplace.

Then, there was Kant. Kant said that space was in the mind, but that the properties of space were the same for all human beings. According to Kant, space was the unique form of all possible perceptions. We order all our perceptions in our pure *a priori* intuition of space—and that space was Euclidean.

If we're going to have a non-Euclidean geometry, those arguments need to be answered—especially Kant's. How can we even imagine a 3-dimensional non-Euclidean world? Of course, we can write equations that describe it, but is it possible for us to order our perceptions in a 3-dimensional non-Euclidean space? Mind-bending questions. In my opinion, the most important person to address the key question—can we order our perceptions in a non-Euclidean space?—was a German, Hermann von Helmholtz.

Helmholtz was a pretty amazing guy. He did physics. He's an independent co-discoverer of the law of conservation of energy. He worked on the physiology of perception—if you've heard of the Young-Helmholtz theory of color vision, for instance, Helmholtz is that Helmholtz—and he did psychology. So, Helmholtz is exactly the sort of person who would care about how we imagine or perceive space.

Helmholtz wrote an influential essay called "On the Origin and Significance of Geometrical Axioms." By "origin" he did not mean history, but the psychological underpinnings—and by "significance," he didn't mean the logical implications of the axioms, but what they tell us about the real world.

So first, then, origin. About the origin, Helmholtz asks: "Where do we get the postulates of geometry—postulates which," Helmholtz says, "are unquestionably true, yet incapable of proof, in a science where everything else is reasoned conclusion?" "The question: Where do we get the postulates? is hard to answer," Helmholtz says, "because of the readiness," he says, "with which results of everyday experience become mistaken for apparent necessities of thought."

So, the only way to really understand what's going on is to divest ourselves of this 2000 years of Euclidean experience. To help us do this, Helmholtz starts by imagining a simpler example of how geometry might be invented. "Suppose," he says, "there were intelligent 2-dimensional beings living on the surface of a sphere. At first, they don't know it's a sphere. But they look around them, and they invent a geometry. They know that they're on a surface, and they have no idea of the possibility of moving into a third dimension—any more," says Helmholtz, "than a person born blind could have an idea of color."

These 2-dimensional beings who live on the sphere would, of course, want to find the shortest distance between 2 points on their surface—what we've done repeatedly in this course by stretching a string. If these beings lived not on a sphere, but on a plane, they'd invent Euclid's geometry.

They'd say, only one straight line can be drawn between 2 points; a straight line can be extended as long as we like in either direction, and the ends will never come together; given a line and a point not on the line, there is only one parallel to the line through that point, and so on. But if they lived on the sphere, that is not the geometry they would invent, says Helmholtz.

On the sphere, their line—the shortest distance between 2 points—would be the shorter arc of the great circle passing through those 2 points. The notion of parallel lines? That wouldn't even occur to them. Every pair of lines intersects—twice. The sum of the angles of a triangle is greater than 2 right angles. It isn't even always the same, by the way—the bigger the triangle, the larger the sum of the angles. We can see this intuitively. A very tiny triangle on a sphere is very close to being a plane triangle, so the sum of its angles is close to 2 right angles. But when we get a really large triangle—like from the North Pole, one vertex at the North Pole and the other 2 along the equator—we can easily get the sum of its angles up to 3 right angles.

So, on the sphere, there aren't any similar triangles—none, because the bigger the triangle, the greater the sum of the angles. So, no smaller triangle can have the same shape. These 2-dimensional beings on the sphere aren't stupid. But still, they'll invent a different system of geometry than would 2-dimensional beings who live on a plane.

Helmholtz then goes on with another example. Suppose these 2-dimensional beings lived on the surface of what the 19th-century geometer Eugenio Beltrami called a "pseudosphere." The pseudosphere has a geometry much like the saddle surface we looked at when we talked about Lobachevskian geometry. Here, the shortest distances—I'll call them lines—are, like the red lines in the picture, infinitely extendable. That's like the plane. But unlike the case in the plane, given a line and a point not on the line, there's more than one line through that point parallel to that is not intersecting with the given line. "So," says Helmholtz, "the 2-dimensional beings who live on this surface would invent exactly the geometry invented by Lobachevsky."

We, unlike these creatures he's talking about, live in 3 dimensions. So it's easy for us to see what's going on in each of these different 2-dimensional worlds. But what about for spaces of 3 dimensions? "So far," Helmholtz says, "our experience in 3-dimensional space has been Euclidean." But must it be? We can, of course, write down equations for 3-dimensional non-Euclidean geometry. But can we represent to ourselves—can we imagine—a space of another sort than the Euclidean, a 3-dimensional space of another sort than the Euclidean? Can we order our perceptions in such a 3-dimensional non-Euclidean space? If we can do that, Kant is wrong. If Kant is wrong, the postulates of geometry are not necessary consequences of an *a priori* transcendental form of intuition.

So, let's see if, through new experiences, we can learn to order our perceptions in a non-Euclidean 3-dimensional space. Helmholtz says: "Yes, we can." Look at the reflections in a convex mirror. You've probably got one of these on your car. There's a warning printed on every convex mirror on a car. You know what it says—or rather what it means. It means—Warning: The space you see in this mirror is not Euclidean.

This well-known picture by M. C. Escher that we've already looked at illustrates this alternate 3-dimensional non-Euclidean reality. Notice in particular how the parallel top and bottom of the bookcase are not equidistant.

Here's another example, this one in a real convex mirror. If you look at your own reflection in a convex mirror, you'll see something like that. Remember the artist Parmigianino and his self-portrait? Parmigianino's hand looks too big, right? Once I was teaching about this, and I passed a convex mirror around the class, and one of the students looked at himself in the mirror, and he said: "Hey, this is distorted." I said: "You're a Euclidean chauvinist." That's a strong way of putting it, but it does make the point. We are used to Euclidean perceptions. This doesn't mean that there can't be others.

So, here's a non-Euclidean space. Can we learn to order our perceptions in it? If you've made it safely so far using your car's convex mirror, the answer "yes" is pretty plausible. You don't look in that convex mirror and you say: "Gosh, that car looks very far away, so it's perfectly safe to pull out into this lane." You have had enough experience to know where the car is. So, you can order your perceptions in a non-Euclidean 3-dimensional space. It just takes practice.

Let me illustrate this with a story. Once I was waiting for a plane in the Denver airport for a very, very long time, and I was sitting right near the exit ramp from one of the gates. The ramp bent 90 degrees about halfway down, and there was a large convex mirror at the bend, so that people coming up or down this corridor could see the traffic coming around the corner.

I didn't have anything else to do, so I amused myself by looking down the corridor at the convex mirror and predicting what I was going to see when the people actually came around the corner to the part of the ramp where I could see them for real. How tall would they be? How fast would they be going? What shape would the things they were carrying actually have? And so on. I got pretty good at it in the 2 hours that I was sitting there. So maybe Kant is wrong. It looks as though we can learn to order our perceptions in a non-Euclidean space.

"But," you might say, "it's a mirror. So it's just an illusion, the world in the mirror. Only our Euclidean world is real." "Now," Helmholtz asks, "are you really that sure? How can you tell which world is the real one, and which one is just the reflection of the real world in some kind of funhouse mirror?" Helmholtz asks you: "Imagine that you're having a dialogue with the person in the mirror." So you say: "My world is real—yours is distorted." He says: "How can you tell?" You say: "Well, in your world, when you get closer to the mirror, you grow bigger—and when you get farther away from the mirror, you get smaller. This violates the obvious fact that when you move something around through space, it stays the same size and shape." He says: "Really? Gosh. Let me try that." So he takes a yardstick. He comes up near the mirror, near the interface between his world and ours, and he measures himself. "I'm 6 feet tall," he says. Then, he goes off to the back of the room, far away from the mirror, and he measures himself again. "No, you're wrong," he says, "I'm still 6 feet tall." You say: "Oh, yes, but your yardstick changed its size also!" He says: "Come on, a likely story!"

"So," Helmholtz says, "there is no geometrical experiment that you can do that will decide the question of which one of these worlds is the real one. We can't tell by pure geometry whether 2 bodies that are the same relative to each other are always really the same, or whether they've both changed in exactly the same proportion."

If this sounds like relativity theory a little, that's no accident. This is a kind of geometrical relativity—there's no geometric experiment you can do to tell these 2 worlds apart. There's a straight intellectual line of historical influence in German philosophy—from Helmholtz on, ultimately, to Einstein.

Helmholtz isn't done with this example, either. He says: "There is no geometrical way of telling which world is real. But the physics of these worlds," he says, "is different." Newton's first law, for instance. In the mirror, does every object in motion—say, moving towards us in the mirror—if acted on by no force, continue to move with a constant velocity in a straight line?

No, it speeds up as it approaches the mirror. Its velocity depends on where it is. Is that true in the real world, of bodies moving with no forces acting on them? Maybe not—Newton's physics says no. But the key point isn't whether it's yes or no for Helmholtz; the key point is this: Whether this is true or false in the real world is not a question about the intuitions that always exist in the human mind—it's an empirical question.

So you can call our minds' ideas of space intuition if you want, but this isn't Kant's idea of intuition, something that is *a priori*. For Helmholtz, our idea of space is empirically based. It's based on our experience of living in the world, just as it was for his 2-dimensional beings. Our idea of space is gained through sight and through touch. For Helmholtz, Kant to the contrary, our ideas about space are not unique, let alone necessarily Euclidean—so, Helmholtz in Germany.

Another interesting perspective on these questions comes from France—from the great French mathematician, physicist, and philosopher of science: Henri Poincaré. Poincaré disagreed both with Kant and with Helmholtz. "If, as Helmholtz says, geometry were an experimental science," Poincaré says, "then geometry wouldn't be an exact science. It would be subjected to continual revision."

But Poincaré is on the other side of non-Euclidean geometry, so he knows that we don't have just one space in our minds. So, he says: "The geometrical axioms are neither synthetic *a priori* intuitions (that's what Kant said) nor experimental facts (that's what Helmholtz said). Geometrical axioms," says Poincaré, "are conventions." Then, how should we decide which set of axioms to use—Euclidean, Lobachevskian, or Riemannian? Poincaré says: "Our choice among all possible conventions may be guided by experience, but our choice remains free"—as long as we avoid contradictions.

So, the axioms of geometry are really only "definitions in disguise." Poincaré concludes: "So what are we to think of the question: Is Euclidean geometry true? Is Euclidean geometry true? The question has no meaning. We might as well ask if the metric system is true, and the old weights and measures are false. One geometry can't be more true than another; it can only be more convenient."

Now to England. In the English-speaking world, a seminal thinker about geometry was the mathematician and philosopher W. K. Clifford. Clifford was inspired by Helmholtz's philosophy of geometry, but Clifford put it all in the broadest possible historical and philosophical context. Clifford said—and what he said is a historical generalization that we've seen throughout this course—Clifford said: "It used to be that the aim of every scientific student of every subject was to bring his knowledge of that subject into a form as perfect as that which geometry had attained."

But no more. "What Copernicus was to Ptolemy," Clifford wrote, "so was Lobachevsky to Euclid. Before Copernicus," said Clifford, "people thought they knew all about the whole universe. No more," says Clifford, "now we know that we know only one small piece of the universe, our own little solar system—likewise with geometry."

See, before non-Euclidean geometry, the laws of space and motion implied an infinite space and an infinite time, whose properties were always the same—so we knew what is infinitely far away just as well as we knew the geometry in this room. "Lobachevsky has taken this away from us," said

Clifford, "sure, space is flat and continuous—but this is true just as far as we can explore, and no farther."

Clifford is very careful about that phrase "just as far as we can explore." Clifford says: "Space is always flat—within the error of measurement." It's amazing to me, by the way, to find statistical reasoning creeping into Clifford's discussion of geometry, but it's supremely relevant.

You think something is flat? "The smoothest polished surface that can be made," says Clifford, "is precisely the one most completely covered with little tiny ruts and furrows. All we know for sure," Clifford says, "is that any deviation from flatness is too fine-grained for us to perceive its discontinuities, if it has any."

Clifford says: "If the property of elementary flatness exists on the average, the deviations from it being too small for us to perceive, we would have exactly the conceptions of space that we have now." Let me put it another way than Clifford does: Space is statistically flat.

Clifford uses these ideas to shoot down the Newtonian philosophy of science— that is, to shoot down the idea that we can have a universal theory of gravitation that applies to all bodies whatsoever. Clifford writes: "Between an inconceivably small error and no error at all, there is … an enormous gulf. A law is, practically, universal when it is more exact than experiment for all cases that might be got at by such experiments as we can make." That's all we have, ladies and gentlemen.

Within our own measurement error—to say, as Newton said, that something is "true of all cases whatever"—we can never have that. So, Clifford uses non-Euclidean geometry to come up with the conclusions we saw earlier coming from people who did statistics and probability.

Some influential thinkers still stood by Euclid in the 19th century. For instance, the famous English economist William Stanley Jevons said: "No, transcendental or necessary truth is not produced by experience; it is recognized rather than learned." We hear the sort of psychological echo of Plato there.

The great English algebraist Arthur Cayley wasn't at all convinced by Helmholtz's 2-dimensional beings on the sphere. Cayley said: "Those beings would [in fact] invent 3-dimensional Euclidean geometry. It would be a true system, applied to an ideal space, not necessarily the space of their experience." The Dutch philosopher J. P. N. Land thought that Helmholtz's convex mirror experiment proved nothing. Land said: "The world in the mirror requires practice to interpret in a Euclidean way," but we can learn to do it.

But Clifford's views seemed to fit better with the increasingly empirical nature of 19th-century natural science. Popularizers of natural science—like the physicist John Tyndall and the Darwinian paleontologist Thomas Henry Huxley—insisted that people can only claim to know the information received through the senses. "Transcendental realities—listen here, Plato and Kant—transcendental realities," say Tyndall and Huxley, "are both unknown and unknowable."

Because of non-Euclidean geometries, and also because of the non-commutative algebra of Hamilton, the prevailing philosophies of mathematics and science have fundamentally changed since Plato and Aristotle.

Plato said that the world of sense experience was an imperfect model of the realm of the forms. Aristotle thought that the objects of mathematics were idealized abstractions from the world of sense experience. But in both cases, both for Aristotle and Plato, the objects of mathematics formed a unique and consistent system with a unique relationship to the world.

In the 19th century, it turns out that there are many different possible mathematical models for reality, of which at most one can be correct. Some of these mathematical structures have been invented with no reference at all to the world of sense experience.

Reacting to all of this, the philosopher Bertrand Russell said: "Mathematics is the subject where we never know what we're talking about, or whether what we're saying is true." Or, in the words of the first great American mathematician, Benjamin Peirce "Mathematics is the science that draws necessary conclusions." So, we keep the idea of logical structure that we got from the Greeks, but we now have a radically different idea about the subject matter of mathematics. The subject matter of mathematics now is—anything that happens to satisfy our basic axioms, and the basic axioms, as long as they are mutually consistent, can be anything we like.

Now, let's move from the logical to the psychological. When we do this, we'll find equally fundamental changes in the way scholars talk about mathematics. Even before non-Euclidean geometry, people like Bishop Berkeley and Thomas Reid had pointed out that we don't really see distance—we merely infer distance from the angles that we do see. Here's a striking example of what they meant. Look at the corner of a room, where the ceiling and 2 walls meet. When we see such corners in the physical world, we perceive this as a place where 3 90-degree angles come together. But if you measure the individual angles as they actually appear and are projected onto the eye, or here on the screen if you measure them—each of those angles is greater than 90 degrees, and together they add up to 360 degrees. Our visual space is not the same as the space we claim to see.

We aren't very good at parallel lines, either. Helmholtz did an experiment where he asked people in a dark room to put little points of light that got progressively farther away into 2 lines that always maintained the same distance from each other—that is, parallel lines. But the supposedly parallel lines that these people made out of these points of light turned out to curve away from the observer, not to be straight parallel lines at all. Experiments like this, many experiments like this, have led psychologists to conclude that visual space—the space we see—isn't represented by any consistent geometry.

Cultures other than the Western also have alternatives to Euclideanness. Different cultures speak about space quite differently. The cultural linguist Stephen Levinson has explicitly asked: "Is Kant right? That is, are spatial categories in language direct projections of humanity's shared innate conceptual categories?"

"There's no evidence for this," Levinson says, "and considerable evidence that suggests otherwise. In both language and concepts, people in other cultures have different ways of ordering their perceptions than in Euclidean space."

For instance, particular directions may have special connotations, and closeness can be cultural as well as metrical. The anthropologist Edward Evans-Pritchard tells us that, among the Nuer people of the southern Sudan, a Nuer tribe 40 miles from another Nuer tribe is considered closer to it than a Dinka tribe 20 miles away. We have a related concept in Los Angeles, by the way: Places that are far distant but quick to get to are said to be "freeway close."

Some cultures use the idea of a fixed coordinate system—having 4 cardinal directions like north, south, east, west, and referring locations to those, as we do when we say: "The house is north of the tree." But other cultures are more like Leibniz; they recognize only the relations between different bodies, as when you say: "The house is to the left of the tree." For that, you don't need an outside coordinate system.

There is yet another way of ordering our perceptions, where the intrinsic properties of the object define the spatial location, as when you say: "The house is on the mossy side of the tree." Some cultures strongly emphasize only one of these methods of ordering objects in space—while others, like ours, use a variety of these methods.

One illustration of these different ways of ordering our perceptions in space in our own society is the way we give directions. Some people give kind of a Newtonian way of locating objects in space. They'll give you directions saying: "Go north for 5 miles, and then turn east for 2 miles." But I know somebody who says: "I can't follow that kind of directions." She prefers directions like these: "Turn left at the stop sign. Keep going straight until you get to the shopping center, and then turn right." Her way is also the way GPS systems give directions—and that's Leibniz's relational view of space.

In fact, GPS navigational systems are changing people's supposedly innate intuitions of space. I had a Maryland cabdriver tell me this. He said: I used to have the whole geography of greater Baltimore in my head. I always knew where I was in relation to that framework. I don't any more now that I have the GPS. Now I think about each trip as, drive straight until such-and-such exit, then make 2 right turns, then make one left turn at such-and-such landmark. When I leave off my passenger, and I want to get back to the expressway, I just reverse that—instead of 2 right turns, one left turn, I do 2 left turns, one right turn. It's really different.

The psychological, linguistic, and cross-cultural evidence is another challenge to Kant's views of space as the form of all possible perceptions common to all human minds. That challenge is in many ways just as compelling as the challenge coming from the invention of non-Euclidean geometries.

So, questions about geometry are deeply involved in major modern controversies—not only about philosophy, but also about psychology and about the alternatives to the Western picture of the world. These questions also affect how artists see the world. In the very next lecture, we'll see how that happened.

# Lecture Thirty-Three
# Art, Philosophy, and Non-Euclidean Geometry

**Scope:** We argued in Lecture Twenty-Nine that artists helped shape the view that space is Euclidean, and that geometry helped the artists produce their art. We have given many examples of how ideas related to Euclidean geometry shaped philosophy. A change in geometry, we might well conclude, ought to affect philosophy and art. Accordingly, in this lecture, we will describe the impact of non-Euclidean geometry (and to some extent, of the theory of relativity) on philosophy and the visual arts. We will see how multiple geometries and relativity theory helped to break down absolutist philosophies and even to promote multiculturalism. We will also look at examples of innovation in the visual arts whose practitioners claimed to be carrying out a non-Euclidean agenda. We may conclude that, whether as motivation or merely as metaphor, non-Euclidean geometry has aided in dispelling the authority of Enlightenment philosophy.

# Outline

I. Now that we have introduced some of the philosophical issues raised by non-Euclidean geometry, we turn to what modern thinkers, and modern artists, have made of it.

   A. After relativity, the physicist Sir Arthur Eddington pointed out that "to free our thought from the fetters of space and time is an aspiration of the poet and the mystic"—but now, it is the physicists who have made this happen.

   B. We look at an important literary intellectual, José Ortega y Gasset, as representative of those who used non-Euclidean geometry as one argument for attacking the dogmatisms of the past.
      1. For Ortega, one such dogmatism is seeking absolute truth.
      2. Another is provincialism: universalizing how it is in your neighborhood (or culture).
      3. The third and fourth dogmatisms are linked: utopianism and rationalism: trying to reduce the world to principles of pure reason.

**C.** Gaston Bachelard, author of part of the surrealist attack on reason and logic, cites Lobachevsky's non-Euclidean geometry as a source while he advocates restoring reason to its true function: "a turbulent aggression."

**D.** P. D. Ouspensky, author of *Tertium Organum*, even attacks conventional logic, which he wants to transcend:

    **1.** "A is both A and not-A."

    **2.** "Everything is both A and not-A."

**E.** Finally, the French cultural theorist Jean Baudrillard applied the same ideas to shoot down what was left of the 18th-century ideas of progress.

**II.** Modern artists and theorists of art were excited by the idea of new geometries, both for their artistic possibilities and as attacks on the older, restrictive views of mathematics and thought.

**A.** There is some conflation of non-Euclidean geometry and the fourth dimension; both are seen as an attack on Euclidean norms.

    **1.** Even those who dealt only with the fourth dimension and not with non-Euclidean geometries owe much to non-Euclidean geometry, since that is what prepared the way for the acceptance of other alternative kinds of space.

    **2.** The key point, after all, was a new freedom from the tyranny of established laws.

**B.** Edwin Abbott's *Flatland* and psychologist G. T. Fechner's discussion of 2-dimensional creatures as shadows are provocative examples of modern theories of the fourth dimension.

**C.** Maurice Princet challenged the cubists to reverse the prejudices of Renaissance perspective.

**D.** There are many other attacks on the idea of Euclidean space and on traditional perspective.

    **1.** Jean Metzinger and Albert Gleizes say that Riemann allows the artist the freedom to deform objects in space.

    **2.** There is also the artistic love of paradox. Tristan Tzara asserted that a "clash of parallel lines" is possible because the artist can transcend contradictions.

    **3.** Modern artists count on the existence of a higher reality, which artists alone can intuit and reveal.

**III.** We look at various works of art that illustrate the foregoing points.

    **A.** We look at the multiple viewpoints in some cubist masterpieces:
  1. Gleizes, *Landscape.*
  2. Juan Gris, *Guitar and Flowers.*
  3. Pablo Picasso, *Portrait of Ambroise Vollard.*
  4. Picasso, *Still Life with Chair Cane.*
  5. Georges Braque, *Portugais (the Emigrant).*
  6. Braque, *The Clarinet.*
  7. Picasso, *Three Musicians.*

    **B.** We look at some artists playing with the fourth dimension.
  1. Marcel Duchamp, *The Bride Stripped Bare by Her Bachelors, Even: (The Large Glass).*
  2. Duchamp, *To Be Looked at from the Other Side of the Glass, with One Eye, Close to, for Almost an Hour.*
  3. Tony Robbin, *Drawing #47.*

    **C.** We look at sculptures by Pevsner, Man Ray, and Hepworth inspired by physical mathematical models.

    **D.** We look at some artists dealing with parallels, curved space, deformable lines and shapes, and curvature of space-time.
  1. René Magritte, *Where Euclid Walked.*
  2. Duchamp, *3 Standard Stoppages.*
  3. Salvador Dalí, *The Persistence of Memory.*
  4. Max Ernst, *Euclid.*
  5. Yves Tanguy, *Indefined Divisibility.*

**IV.** In the $20^{th}$ century, revolutionary ideas based on geometry had great impact on well-known writers and artists and shaped how they thought about their art.

**Essential Reading:**

Henderson, *The Fourth Dimension and Non-Euclidean Geometry.*

Kline, *Mathematics in Western Culture*, chap. 26.

**Suggested Reading:**

Kemp, *The Science of Art.*

Miller, *Einstein, Picasso.*

Ortega y Gasset, "The Historical Significance of the Theory of Einstein."

Robbin, *Shadows of Reality.*

**Questions to Consider:**

1.  Some artists have geometry in mind but get the geometry wrong; thus there is nothing non-Euclidean about Tanguy's painting of parallels, Ernst's non-Euclidean fly, or many of these pictures of the fourth dimension. Does it matter?

2.  The art historian H. W. Janson has written, "Collage cubism [like Picasso's *Still Life with Chair Cane* and *Three Musicians* and Braque's *Le Courrier*] offers a basically new space concept, the first since Masaccio." That is quite a claim. Do you agree?

# Lecture Thirty-Three—Transcript
## Art, Philosophy, and Non-Euclidean Geometry

Here we go again. Geometry is changing the way people look at the world—for the third time. The first time was when the Greeks invented proof. The second time was in the Renaissance, when the artists used the geometry of perspective.

As for the third time—well, let's listen to the architect Zaha Hadid. She's the first woman, by the way, to win the Pritzker Architecture Prize. Let's hear Zaha Hadid describe her take on the world view of the 21st century.

She says: "The most important thing is motion, the flux of things, a non-Euclidean geometry in which nothing repeats itself, a new order of space." Here's an example of her work. That's the interior of the Bridge Pavilian in Zaragoza in Spain. Hey, we aren't in the Renaissance "Ideal City" any more.

In the previous lecture, I introduced some of the philosophical issues raised by non-Euclidean geometry. Now we'll see what some 20th-century artists and literary theorists have made of non-Euclidean geometry. We'll see how non-Euclidean geometry has reshaped—literally reshaped—both our artistic and our intellectual landscapes.

First, just a little bit more intellectual background. As we saw in the previous lecture, in France, the great mathematician and physicist Henri Poincaré said that: "Which geometry we choose is solely a matter of convention." That is, Poincaré says: "It's like choosing which language we speak or how we dress. The axioms we choose for geometry," Poincaré said, "are neither synthetic *a priori* judgments [as Kant would say] nor experimental facts [as, say, Fourier or Helmholtz would say]. The axioms are conventions. We have to avoid contradictions, but otherwise, our choice of axioms is free." That statement by Poincaré about freedom of choice in geometry was very attractive to a number of artists, especially in France.

Also in the direction of freedom, the English physicist Sir Arthur Eddington reflected on relativity theory and the new geometries. Eddington wrote: "To free our thought from the fetters of space and time is an aspiration of the poet and the mystic. But now," said Eddington, "it's the mathematicians and the scientists who have made it happen." Eddington's got something there. So, let's see what the effects on literary and social theory have been.

One especially influential thinker was the Spaniard José Ortega y Gasset. Ortega used both non-Euclidean geometry and relativity to shoot down 4 of what he called the "dogmatisms of the past." The first dogmatism is "absolutism." Ortega said: "All absolutisms are wrong, whether in geometry, physics, or philosophy. Reality," Ortega says, "is relative."

Second dogmatism: "provincialism." Assuming that our own experience or values are universal—that's just wrong. That's provincialism, universalizing how it is in your neighborhood. "Euclidean geometry, [this is going to sound like W. K. Clifford]" Ortega says, "was provincial. It was an unwarranted extrapolation of what was locally observed, extrapolated, to the whole universe. Instead," says Ortega, "reality organizes itself to be visible from all viewpoints." If, to those of you familiar with Cubist art, that sentence sounds like Cubist art, that's no accident—as we will soon see. Ortega goes on: "Einstein's theory of relativity, which requires new geometries of space-time, promotes the harmonious multiplicity of all possible points of view.

Now that's not just for physics," says Ortega, "this also implies that non-European cultures have valid points of view." Ortega says: "There is a Chinese perspective that is fully as justified as the Western."

The third and fourth dogmatisms that Ortega talks about are linked—he calls them "utopianism" and "rationalism." See Plato and Descartes being shot down here. "Both utopianism and rationalism are wrong," says Ortega. "Since the Greeks," he says, "reason has tried to build an idealized world and say, this is true, this is how it is." Before relativity theory, Ortega claims, the physicist Hendrik Lorentz has said: "Matter must get smaller as it goes faster—that is, matter yields—so that the old laws of physics can continue to hold."

"Not so," says Ortega, "listen to Einstein," says Ortega. Einstein says: "Space yields: geometry must yield, space itself must curve." That's how things are. Ortega interprets these 2 different views politically. Ortega says: "Lorentz might say, 'nations may perish, but we will keep our principles.' But Einstein would say," according to Ortega, " 'we must look for such principles as will preserve nations, because that's what principles are.'" Very, very interesting. Ortega was not the only person to try to do this kind of social criticism coming out of their interpretation of geometry and relativity.

For instance, the surrealist theorist Gaston Bachelard, a Frenchman, wrote an essay attacking both reason and logic. Bachelard cited Lobachevskian geometry as a source when he advocated restoring reason to its true function. What is the true function of reason, according to Bachelard? I'll tell you what it's not. It's not to shore up the agreed-upon order. "As the new geometries show," says Bachelard, "reason is a turbulent aggression." Wow.

The Russian P. D. Ouspensky was influenced by these same ideas. Ouspensky attacked the limitations of 3-dimensional space. He identified the limitations of 3-dimensional space with Euclidean geometry. But Ouspensky went even farther. Now Aristotle is in somebody's sights. Ouspensky went even farther and attacked the basic principle of all of logic in something called, by the way, [*Tertium Organum*] (*The Third Organon*). Remember the *Organon* of Aristotle, and then the new *Organon* of Francis Bacon? Okay, this is *The Third Organon*. So, you may think that a statement can't be both true and false. You remember "All A is B" and all that stuff. Ouspensky declares: "A is both A and Not-A"—and even more: "Everything is both A and Not-A." So much for Aristotle and deductive logic.

Finally, the last person of this sort that I'm going to quote: the modern French cultural theorist Jean Baudrillard. He applied these same ideas to shoot down what was left of the 18th-century idea of progress. He's more or less a contemporary of ours. Here's what Baudrillard wrote: "In the Euclidean space of history, the fastest route from one point to another is a straight line, the one of Progress and Democracy." You remember Condorcet, right? "This however," he goes on, "only pertains to the linear space of the Enlightenment. In our non-Euclidean space of the end of the 20th century, a malevolent curvature invincibly reroutes all trajectories." This is just amazing.

All right for literary theory and social theory. Now for the visual art itself. Modern artists and theorists of art (theorists of modern art) were also very excited by the idea of new geometries. They sometimes even get the geometry wrong, but they still have a tremendous amount of fun with it. Artists often equated non-Euclidean geometry with the fourth dimension— why should we be limited to just 3 dimensions? The way the artists saw it, both non-Euclidean geometry and the fourth dimension seemed to attack the conventional Euclidean norms. But even Euclidean ideas about the fourth dimension (because you could have a 4-dimension Euclidean geometry) owe a debt to non-Euclidean geometry historically, because non-Euclidean geometry prepared the way for conceiving alternative kinds of space. You remember Réaumur? The key point for the arts—after all, from all of this— was a new freedom from the tyranny of established laws.

Just a little bit more theory, and then we'll be ready to see how all these ideas are reflected in actual works of visual art. First, there's the theory of the fourth dimension. There's a famous book written by Edwin Abbott, written in 1884, popularizing the fourth dimension. The book is called *Flatland*. In the country of Flatland, everybody lives on a 2-dimensional plane. (I think that there's probably influence on Abbott by Helmholtz, but that's by the way.) These 2-dimensional beings live in a 2-dimensional plane, and then they're visited by a sphere. The sphere comes from the third dimension. But the Flatlanders can't conceive of the sphere. This sphere comes along, and when it touches the plane that these guys live in, they see a circle—and so the 2-dimensional beings on the plane interpret the sphere's intersections with their plane as merely a succession of circles. The problem that the Flatlanders have in imagining the third dimension, according to Abbott, is the same as the problem that we, as 3-dimensional creatures, have in imaging or understanding the fourth dimension. That's a helpful analogy if you want to think about the geometrical fourth dimension.

There's an analogous discussion from a psychologist of perception, the German Gustav Fechner. Fechner was very interested in the psychology of perception all kinds of ways. There's a Fechner's law of perception involving logarithms. It still works for the way we hear the loudness of sound. Fechner said: "You can think of these 2-dimensional creatures as shadows of 3-dimensional figures. In the same way, then," says Fechner, "our world is a 3-dimensional shadow of the fourth-dimensional reality." Sounds a little like Plato, doesn't it? So how can we get at that fourth dimension? Artists say: "We can do it!"

Furthermore, some theorists of art attack the whole enterprise of Renaissance perspective. They say: "Hey, Renaissance geometry doesn't produce realistic painting!" For instance, the French critic Maurice Princet challenged artists: "Let's reverse the prejudices of Renaissance perspective! Instead of portraying objects on your canvas as deformed by perspective, they ought to be expressed as a type." In Renaissance perspective art, a rectangular thing, like this piece of paper, appears on the canvas shaped like a trapezoid. To prove this, I ask that you trace the way this shape, rectangular shape, actually appears on your flat screen, or the table. You'll see if you look at how it appears on your flat screen, it's not a rectangle—it's a trapezoid. "A table shouldn't look like a trapezoid," says Princet, "it ought to be straightened out into a true rectangle." Likewise, if you look at a glass, a water glass, the oval portrayed in perspective should become a perfect circle—and that's done in Cubist art. For instance, here is Jean Metzinger.

Look especially at the teacup that she has—the top view and the side view not distorted by perspective—and you also might want to look at her face and at her hands. So this is our first artistic example of this reversing the prejudices in Renaissance perspective.

Also, in modern art, there's a new approach to treating objects in motion. In Euclidean geometry, as you'll remember, when you move something, it keeps its shape and size. But theorists of Cubism, like Jean Metzinger and Albert Gleizes, they say: "Riemann's geometry gives artists the freedom to deform objects in space." Pretty soon, we'll see Cubists and Surrealists doing this.

Finally, in modern art, there's a love of paradox. For instance, Tristan Tzara, the founder of the art movement called "Dada," speaks of "the precise clash of parallel lines." Does that sound like a contradiction—"precise clash of parallel lines"? "Don't worry," says Tzara, "the artist can transcend contradictions."

So, to sum up what all these theorists were saying: Beyond the Euclidean world that conventional people think they inhabit, there is a higher reality—a reality that artists alone can intuit and reveal. Non-Euclidean geometry, then, both liberates and legitimates these new approaches to art. Of course, there are many other influences on the new art than just geometry. But just for this lecture, I want to focus on the new geometries so you can appreciate their roles in helping inspire some masterpieces of modern art.

First, we'll look together at the multiple viewpoints of some masterpieces of Cubism. The objects are not "deformed by perspective" but instead "expressed as a type." For instance, here's a landscape by Albert Gleizes.

We'll look at that a little bit. Here is a painting by Juan Gris, *Guitar and Flowers*. In this picture, we see multiple views of the same subject united into one picture—just as his fellow Spaniard, Ortega y Gasset, had advocated.

Here's a reasonably well-known example. This is by Pablo Picasso, another Spaniard: *Portrait of Ambroise Vollard*. Again, you have these planes and angles from multiple points of view. This man is not placed in a visually graspable 3-dimensional space.

Here's Picasso again: *Still Life with Chair Cane*. It's a collage. This is the beginning of what is now called "Collage Cubism." Of course, nowadays every third grader knows how to cut up magazines and make a collage. This pioneering work by Picasso is a paste-up, and the caning actually comes from a chair. Collage Cubism features cut and pasted flat pieces of material, and a collage of such materials. They represent the image, but they also remain themselves. Again, like other Cubist art, we have multiple points of view.

Here's another Cubist masterpiece. This is by Georges Braque. The painting is called *The Emigrant*—again, multiple points of view. This one also looks as though it's reflections in the fragments of a broken mirror. It's an interesting image for multiple points of view.

Here's another Braque: *The Clarinet*. Now you are getting both the color and the texture of the object at the same time that the object is being represented by abstract planes in different materials.

Now a masterpiece of this particular approach to art. This is Picasso—*The Three Musicians*. I think this is just absolutely wonderful. Go and see it in the Museum of Modern Art if you're ever in New York, and imagine—as we imagined when we looked at Masaccio's *Trinity* in the lecture on the Renaissance—imagine what it meant to look at this painting by Picasso for the first time. The separate pieces of *The Three Musicians* fit together perfectly, like blocks in architecture, and yet they also create this marvelous whole image. It's even hard to tell whether this is a paste-up or a painting.

All right, now let's turn to something else. Let's look at a few artists playing with the fourth dimension. First, this is Marcel Duchamp, and it's called *The Bride Stripped Bare by Her Bachelors, Even: (The Large Glass)*. Duchamp's own notes on this work refer to Henri Poincaré talking about multiple dimensions. So, Duchamp claims that *The Bride* is the projection of a 4-dimensional object.

Here's another Duchamp. This one is called *To Be Looked at from the Other Side of the Glass, with One Eye, Close to, for Almost an Hour*. Part of the point of this work is to make fun of Renaissance one-point perspective. It's not only the construction of the painting geometrically that does that, but also the title.

Here is a 21$^{st}$-century mathematician and artist called Tony Robbin—mathematician and artist. What Robbin has here are independent planes and solids that really seem, as you look at them, to come from 4-dimensional space.

Let's look at a different kind of non-Euclidean geometry influence on art. You remember the physical model—the saddle-shape, the hyperbolic paraboloid—the physical model of a Lobachevskian surface that we saw back in Lecture Thirty? Such physical models of mathematical surfaces were used by mathematicians to teach about such surfaces and to study the geometry of such surfaces. But seeing such surfaces also inspired some artists. For instance, here's the constructivist artist Antoine Pevsner. This is a sculpture called *World*. I think you can see the influence of models like that.

Here's another piece of sculpture. This next one is also by Pevsner, whose title makes clear its mathematical antecedents: *Spatial Construction in the Third and Fourth Dimensions*. This is lovely, by the way.

Here is another artist who did this: Man Ray. This is called *Lampshade*. Man Ray, in this and other sculptures, used shapes based on mathematical models that he saw in a scientific institute of Paris. What makes this really cool—there's a really cool fact about the scientific institute where he saw this—it's the Poincaré Institute. That's the guy I quoted earlier about non-Euclidean geometry and different conventions.

Last—but not least among this particular set of influences of geometry— here is Barbara Hepworth, who was also inspired by looking at various mathematical models. This one is called *Pelagos*. Isn't it beautiful?

Let's look at something else. Let's look at artists dealing with parallels, with curved spaces, with deformable lines and shapes, and with the curvature of space-time. First, on one level this is the simplest of them, and yet it's really marvelously intellectually complex. This is by Rene Magritte, and it's called *Where Euclid Walked*. This work by Magritte has a novel approach to 3-dimensional illusions. Look especially at this repeated shape, which in the picture has 2 different meanings: First, it's the turret of a tower, and then it's the street painted in perspective. While you're looking at the street, enjoy the ironic reference to Euclid in the title of the painting: *Where Euclid Walked*. We're past that now. The title is *Where Euclid Walked*, but we're past that now.

Now we're going to go back to Marcel Duchamp once more. In this painting, imagine those curved lines moving up and down in the painting and being deformed as they go. Duchamp, himself, wrote that these curves and their motion were inspired by Riemannian geometry. But Duchamp is also thinking about probability theory; he referred to this painting as *Canned Chance*.

The next slide that I want to show you is a very famous example of this type of distortion. It's by Salvador Dalí, and it's called *The Persistence of Memory*. Dalí's notes about this painting (Dalí's own notes about what he thought he was doing) refer to non-Euclidean geometry and its curved space-time. You look at those watches, and you think, just say that to yourself: "curved space-time." Also, Dalí contrasted the rigidity of Euclidean structures with those in this painting. The prerequisite for the next remark is that Camembert is a runny French cheese. He called this painting an "extravagant and solitary Camembert of time and space."

Even the titles of some paintings pay homage to non-Euclidean geometry. The symbolist Max Ernst painted a playful work called *Young Man Intrigued by the Flight of a Non-Euclidean Fly*. I did want to show you a Max Ernst, so here is Max Ernst's portrait of *Euclid*. Notice the intersecting planes, and notice the multiple perspectives. Is this a portrait of Euclid? Of Euclidean geometry? Of non-Euclidean geometry? It certainly violates the Euclidean conventions. Again, it's all about freedom. Non-Euclidean geometry, for the artists, is liberating. One more example—well, more than one example, actually, but one additional example now: Euclidean geometry dictates that parallel lines can never meet. Oh yeah? They can if they want to.

So, the surrealist painter Yves Tanguy actually gave us a painting called *Le Rendez-Vous des Parallèles*—(*The Meeting of Parallel Lines*). Tanguy just loved to play with geometric themes. I do want to show you a Tanguy also. Euclidean geometry allows us to divide a line, or plane, or space as many times as we want. So, here is Tanguy's interpretation of that. The painting is called *Indefined Divisibility*—remarkable, really.

Let me summarize what's been going on in this lecture. The American mathematician, Morris Kline, once said: "Non-Euclidean geometry knocked geometry off of its pedestal, but it also set it free to roam." I think that's well said. The freedom that geometers claimed for themselves was magnified greatly when philosophers and artists picked up their ideas. Revolutionary ideas about geometry have shaped thought, art, and architecture. The examples that we've seen in this lecture just scratched the surface. There's a tremendous wealth of these things, and a wealth of theorists of art talking about non-Euclidean geometry as one of the motivating forces in this kind of really amazing art.

I want to close this lecture with an architectural work from my native Los Angeles. This is *Disney Hall* by Frank Gehry. Gehry is at home with all kinds of shapes: positively curved, negatively curved, or Euclidean—whatever he wants to use, he puts it up.

So, revolutionary ideas about geometry are, as this and many of these earlier things illustrated, transforming the world that we live in—not just the world of ideas, but the physical furniture of the world in which we live. Thank you.

# Lecture Thirty-Four
# Culture and Mathematics in Classical China

**Scope:** One way to test our generalizations about the relationship between mathematics and culture is to look at a different mathematical civilization. Chinese mathematics, which was quite sophisticated, developed independently from the Greco-European tradition In premodern China, there was not the close relationship with philosophy that has, since the time of Plato, characterized Western mathematics. We will look briefly at the cultural situation of classical Chinese mathematics. We will focus on the different approach to proof found in China, including the lack of indirect proofs. As an example of Chinese proof technique, we will look at the proof, probably as old as Pythagoras, of the Pythagorean theorem in a classical Chinese text and contrast it with the way Euclid proved the same theorem. We will conclude by asking whether if geometry had been treated less like a branch of philosophy and more like a branch of engineering, Western mathematics might have been less abstract and easier to teach on an elementary level.

# Outline

I.  We have said a lot about the relationship between mathematics and culture in general, and between mathematics and philosophy in particular.

   A.  But we have, so far, confined our discussion to the tradition that began in the Greek-speaking world, a tradition that sought certainty by means of logical proofs about idealized objects.

   B.  The cross-cultural questions we asked about ideas of space can also be posed about mathematics in general.

   C.  Mathematics is found in all cultures and develops differently within them.

   D.  What would a good test case for our generalizations be?
      1.  It would have to be a culture whose mathematics, during the period we look at, developed independently of the Greek tradition.
      2.  It would need to be a culture whose mathematics was complex and sophisticated.

E. An excellent choice is the mathematics of China before major contact with the West.

II. We briefly sketch some major mathematical achievements in the history of mathematics in China: in algebra, number theory, and geometry.

III. We turn to the background of classical Chinese mathematics.

   A. We discuss the cultural setting of classical Chinese mathematics, especially its relationship to engineering problems.

   B. We ask what historians say about the greater interest in technology in China than in Greco-Roman society.

   C. We describe the examination system, its role in Chinese society, and its implications for the learning of mathematics.

   D. We contrast the way philosophy was organized in Greece with that in China.

      1. Greek philosophers competed for students; they emphasized argument and debate to do so.

      2. Rival philosophies in Greece stimulated the twin needs to prove the true and identify the false.

      3. In China, the opposites yin and yang are reciprocal, and their synthesis is emphasized, while in Greece, when there are opposites, there is often confrontation and contradiction.

      4. Although Chinese philosophers argued also, their schools emphasized preserving and transmitting a body of learning where the text had authority.

      5. Greek competitive argument includes disagreement over what form of government is the best, while Chinese philosophy and science often was directed toward the emperor, whose wise and benevolent rule was a social ideal.

IV. We now ask how the Chinese approached the kinds of mathematical questions Euclid considered; in particular, we

   A. We look at right-triangle theory in ancient Pythagorean contrast the Chinese proof of what we theorem with the one given by Euclid... ings in general.

   B. We discuss the way the Chinese... ame pattern in different

      1. Chinese mathematics loc... procedures at a higher level contexts, unifying and of abstraction, but r

**2.** The Chinese, in their mathematical classics, did not use indirect proof in geometry at all. Why?

   **a.** There are linguistic and social theories about this, and we describe them.

   **b.** We challenge these views with the fact that indirect proofs were used in philosophy, even before the coming of Buddhism to China.

**V.** Classical Chinese society and Greek-speaking society were culturally different and produced different, though equally sophisticated, mathematical traditions.

   **A.** There is no single straight-line path along which mathematics must develop and progress.

   **B.** Looking at mathematics in a different cultural setting suggests that there are psychologically different ways of looking at similar mathematical subjects.

   **C.** Knowing this ought to help us do a better job teaching mathematics. Even within the classrooms in one society, we find a variety of different learning styles. None is the best.

### Essential Reading:

Swetz and Kao, *Was Pythagoras Chinese?*

### Suggested Reading:

Bloom, *The Linguistic Shaping of Thought.*

Katz, *A History of Mathematics*, chap. 7 (chap. 5 in the brief edition).

Lloyd, *Adversaries and Authorities.*

### Questions Consider:

**1.** In modern society, calculators and computers can do a great deal of the computation that once made up the bulk of school mathematics, so understanding the logic and abstract ideas of mathematics has become much important. How, if at all, can understanding the relationship between mathematics, culture, and the different ways of thinking conditioned culture improve the way we teach mathematics in schools?

**2.** The Pythagorean theorem was independently discovered in several different cultures besides the Greek: probably in ancient Egypt, and certainly in ancient Babylonia, in classical India, and in China. What is there about that proposition that made it so important to all these different societies?

# Lecture Thirty-Four—Transcript
# Culture and Mathematics in Classical China

Welcome back. I've said a lot about the relationship between mathematics and culture in general, and about the relationship between mathematics and philosophy in particular. But so far, I've confined the discussion to the tradition that began in the Greek-speaking world and moved through the medieval Islamic and Christian worlds into Europe and the Americas. But the cross-cultural questions that I touched on earlier when I was discussing ideas about space can also be asked about mathematics in general. In this lecture, I want to take a look at a tradition in mathematics from outside the West.

Mathematics is found in all cultures. But it has developed differently in different cultures. That, of course, doesn't mean that $2 + 2 = 5$ in some cultures, and equals 4 in others. But it does mean that the mathematical ideas that get explored, the applications of mathematics, and the methods used to establish results—all reflect the needs, and histories, and social organization of the cultures in which the mathematics is created. We've already seen that happening in the Western tradition.

To illustrate—and to test—generalizations like the ones I've just been making, what would be the most illuminating example? First, it ought to be a culture whose mathematics developed largely independently of the Greek tradition. Second, the mathematics in this culture we're using for comparison purposes should be sophisticated enough so that it provides a real test for our generalizations. There is an excellent choice that has both of these properties: the mathematics of China before the 16th century—that is, the mathematics of China before there were major scientific contacts with the West.

So, in this lecture, I'm first going to briefly sketch some of the major mathematical achievements in classical China. It's mostly going to be just a list. It would take too long to explain all the mathematics. But this list will definitely prove the point about the sophistication of classical Chinese mathematics.

Here's the first item on the list: the number system. The Chinese, like the cultures of classical India, like the Islamic world that got it from classical India, and like us—the Chinese had a base-10 number system with place value. The Chinese had recognized negative numbers also already about

2000 years ago and had sophisticated algorithms for working with fractions. So—number systems.

Second, arithmetical and algebraic computations: The Chinese could calculate square roots, cube roots, as accurately as anyone needed. They could solve quadratic equations exactly—and higher-degree equations with any desired accuracy.

In fact, for higher-degree polynomial equations—something like, say, $x^6 - 5x^4 + 7x^3 + 3x = 107$, like that—for higher degree polynomial equations, the Chinese developed an approximation procedure, which—when it was rediscovered hundreds of years later in Europe—became known as Horner's method, after the 19$^{th}$-century mathematician George William Horner. The problems that originally motivated the first Chinese work on polynomials came from finding the areas of irregularly shaped fields, but the general solution methods that the Chinese have go far, far beyond any merely practical problem.

The third area is combinatorics. The Chinese mathematicians knew all about things like "7 choose 3." In fact, the Chinese algebraists solved the problems I mentioned before about polynomials by using combinatoric principles. That is, they knew how to use combinatorics to get what is now called the "binomial theorem"—that's how to raise binomials—that's a thing like $a + b$ to whole-number powers. So, for those of you who like algebra, here is one example: This is $a + b$ raised to the fourth power. Look at the things that aren't $a$'s and $b$'s, the coefficients of each term. The first term, the coefficient is 1; the second term, the coefficient is 4; then 6; then 4; and then 1.

You'll recognize those numbers—1, 4, 6, 4, 1—that we worked out when we were discussing the bell curve. They're "4 choose 0," there's 1; "4 choose 1," which is 4; "4 choose 2," which is 6; "4 choose 3," which is 4; and "4 choose 4," which is 1. What you are doing is counting how many of the factors in each of these products is a $b$—no $b$'s, 1 $b$, 2 $b$'s, 3 $b$'s, 4 $b$'s. By the middle of the 11$^{th}$ century, Chinese mathematicians knew the properties of such combinations, like "4 choose 2," and knew also how they were embodied in what is known in the West as the "Pascal triangle," because Pascal did it. In China, this is sometimes known as the "triangle of Yang Hui" and here's what it looks like. The fifth row, for instance, represents—in Chinese notation—the numbers 1, 4, 6, 4, and 1.

Fourth, geometry: Chinese mathematicians had general results about areas of triangles and other polygons, as well as very accurate approximations to the area of a circle. They also have results about volumes: volumes of pyramids, volumes of spheres, volumes of other solids, including—this result is as early as the 3$^{rd}$ century of the Common Era—the volume of the figure formed when 2 cylinders intersect within the same cube. This is a modern picture of 2 cylinders intersecting.

Feel free to try to calculate this yourself. Without calculus, I think it's very hard. Even with calculus, it's not trivial.

Fifth area in Chinese mathematics: general results about the structure of algebra. The Chinese solved what are called in modern mathematics "systems of simultaneous linear equations." They abstracted from the equations to what modern mathematicians call "matrices and determinants." Also, the Chinese solved what are called "simultaneous congruences." I'll give you a simple example. This is a Chinese example. This is a very simple example, but you'll see what I mean by simultaneous congruences.

The problem is this: If we count a set of things by 3's, the remainder is 2. Let me explain what I mean by that. Let's say you have 11 things—you count 3, 6, 9, and there is 2 left over; that's what it means. We count a set of things by 3, and the remainder is 2. Okay, we're given a set of things, and here's the problem, we count a set of things by 3's—the remainder is 2. If we count them by 5's, the remainder is 3; like for 18, we'd count 5, 10, 15, and then there would be 3 left over for 18; if we count them by 7's, the remainder is 2. Let me run the whole problem by again. If we count a set of things by 3's, the remainder is 2. If we count them by 5's, the remainder is 3. If we count them by 7's, the remainder is 2. How many things are there then in the set?

Feel free to try to solve this on your own. The Chinese had a general method for solving all such problems. It was given in the 13$^{th}$ century by one Qin Jiushao. His method is known in the West as the "Chinese Remainder Theorem." When I took number theory, I studied the Chinese Remainder Theorem. I didn't know the history of it, but that was its name. Oh, you want the answer to the original problem? It's 23.

Last, the Chinese knew what we call the Pythagorean Theorem—and, in fact, made the study of areas central to their geometry, areas rather than lines. In a few minutes, I will show you how the Chinese proved what we call the Pythagorean Theorem. But because their approach to geometry, as you'll see, is so different from that of the Greeks, I want to prepare for explaining those differences by looking at the social and cultural background of the classical Chinese mathematical tradition.

Mathematics in most ancient civilizations, including Egypt and Babylonia (the predecessors of the Greeks), began with practical problems. Of particular importance in China—you have to remember, it was a highly centralized and bureaucratic empire—so of particular importance in China were surveying and engineering, and building canals and fortifications, and computing tax rates and interest rates, and predicting the positions of the sun and the moon to determine the calendar.

The Greeks, especially after Plato and Aristotle, tended to think of pure mathematics as better than applied mathematics, since the goal of pure mathematics was knowledge—rather than just getting something done. Also, as we've seen, Greek mathematics developed hand in hand with philosophy. In China, though, mathematics remained closer to its applications.

There were also differences between Greece and China in the nature of mathematics education and in the audience for mathematical research. Among the Greeks, as we've seen in talking about Plato particularly, mathematics was at the heart of education. For the Chinese, it was just one subject among many.

The Chinese Empire had a civil service selected by a system of competitive examinations. Those examinations were primarily on the Chinese literary classics, although works on mathematics were included, too. The Chinese Empire—besides needing practical geometry—had standardized weights, and measures, and money—so taxation and coining money were especially important tasks for the imperial government.

Studying such useful mathematics, then, and studying them in the classical mathematical texts, was a good way to qualify for the Chinese imperial service. But, studying texts in order to pass a test on how well you've mastered the key passages is not the strongest incentive to create new mathematics. None of this means that Chinese mathem... ans didn't pursue mathematical topics out of theoretical interest. The... It's just that the emphasis often lay elsewhere.

...th mathematics in

Not only was philosophy not intimately interwo...zed quite differently China, philosophical communities in China wer...d for students. They than those in Greece. Greek philosophers ...val Greek philosophers emphasized argument and debate to do... to shoot down what was argued a lot—both to prove what was...na, though, the opposites Yin false. In Greece, where there are op... their synthesis is emphasized. confrontation—that's Plato's diale... and Yang have a reciprocal rela...

Furthermore, although Chinese philosophers, of course, argued with each other, an important goal of their philosophical schools was to preserve and transmit a body of learning, where the text would have authority. In political philosophy, Greek philosophers often debated about what form of government is the best. By contrast, Chinese philosophy and science were often intended for the empire and for the emperor, and the emperor's wise and benevolent rule was a social ideal. So, if we compare classical Greece and classical China, we find a different setting both for mathematics and for philosophy.

Now to the mathematics itself. How did the Chinese approach the kinds of mathematical questions that Euclid considered? Let's first look at the foundational result of right-triangle theory in China, the result that we call the Pythagorean Theorem. Let me start with the Chinese "proof" of the theorem.

The earliest text we have of this proof is from a book called the *Nine Chapters on the Mathematical Art*. It dates from about 200 B.C.E., but the material it describes is clearly much older, and the Chinese version of the theorem may even be contemporary with that of Pythagoras in the 6th century B.C.E.

First, let me explain the basic idea behind the Chinese proof with a modern diagram. We're going to use a right triangle whose smaller leg I'll call a, and whose larger leg I'll call b, and whose hypotenuse—the longest side—is c. Our goal is to prove that the square of the smaller leg, plus the square of the larger leg, is equal to the square of the hypotenuse.

This is a geometric result about the areas of actual squares, even though we are used to writing it algebraically and calling the sides a, b, and c instead of— as the Chinese did—using the word for "leg" for the smaller side, and "thigh" for the larger side (well, we say "leg"; we just don't say "thigh" for the other ne). They use the word for "stretched string" for the hypotenuse hich, by the way, is essentially what "hypotenuse" means. The stretched ng, after all, is how you find the shortest distance between 2 points. Tech rms in geometry sometimes reveal practical origin.

Anyway, let's see take a square whos ea behind the Chinese proof of this theorem. First, side of the square is s what we'd call a + b—that is the length of the Now no matter how w of the length of the 2 legs of the right triangle. same. As you can see as up such a square, its area will always be the of length a + b—therefore at these 2 squares, each of them has a side ot the same area.

We see that the left-hand square is made up of a square whose side is a, and a square whose side is b, and then 4 of the right triangles with sides a, b, and c. We also see that the right-hand square is made up of the square on the hypotenuse and 4 of the same right triangles. If you take away the same thing—4 right triangles—from each of these pictures, what you're left with is a square whose side is a plus a square whose side is b on the left, and a square whose side is c on the right. So those 2 areas—the sum of the squares of the sides of the square of the hypotenuse—must be equal. By the way, I think this is the clearest and most intuitive proof of the Pythagorean Theorem that's around.

Let's look at the actual Chinese diagram from the *Nine Chapters on the Mathematical Art*. It's a little different from the way I presented it. The 4 triangles are found partly inside the $c^2$ figure, with a little square inside whose side is the difference between the 2 legs of the triangle. You have to imagine moving these triangles around, or redrawing the heavy lines as in my first diagram. But conceptually, it is the same as the proof I showed you.

Now that we have an example, let me ask: What are the differences between the Chinese and Greek approaches to proof in geometry? You may have already noticed in the Chinese diagram something we never find in Euclid. In the Chinese diagram, we have a specific numerical example—you can see that the right triangle involved has sides of 3, 4, and 5—although, to be sure, the example stands for the general case, and the proof method does work in general.

Also, the Chinese do not have explicitly stated axioms or postulates, and they do not have precisely stated definitions. I mean, you know what a square is, right? Since the Chinese draw beautiful diagrams—often decorated, by the way, which is something else the Greeks don't do. Look at this one. This is a decorated diagram for a geometry problem. Since the Chinese draw beautiful diagrams, it's always clear what they're talking about. They do have one basic principle. Although they don't state it explicitly as a postulate, it's sometimes stated explicitly in the context of a particular argument, and it is generally presupposed. That principle is, if you dissect an area or volume in 2 different ways, the total area, or the total volume, remains the same. You saw that principle used in the Chinese proof of the Pythagorean Theorem.

Another difference between Chinese and Greek geometry is the way the Chinese geometers identify lines and areas. Chinese isn't an alphabetic language, so they don't label individual points with letters of the alphabet.

Sometimes, they'll label a line with its name—like the word for "leg," they'll write on the right-triangle side. Sometimes, they label an area with a color. If you're familiar with the kind of geometric puzzle called a "Tangram," these originated in China—I think in the 18$^{th}$ century (I'm not sure). But, anyway, you've seen these—you've seen these traditions about area and color combined in another Chinese cultural product.

The Chinese have another thing that's different. Although they know that similar figures have proportional sides, they prefer to express those relationships using areas rather than proportions. Let me give you an example so you see what that means. I'll use modern notation. Here we've got a big rectangle, and this little piece on the top is $a$, and this piece is $c$, and this piece is also $a$, and this piece is also $c$. The sideways piece is $b$; this is $d$—$b$, $d$, and so on. Let's draw the diagonal from the top left to the lower right of this rectangle. The Greek approach would be to say: "Hey, look, we've got 2 parallel lines cut by a third line, so various angles are equal—so the triangles shaped like this, this triangle and this triangle that I've put $x$'s in, they are similar—they have the same shape."

That means, the similarity of the triangle means that the sides are proportional—so $a$ is to $c$ as $b$ is to $d$. The Chinese, with the same diagram, would prefer instead to describe the same relationships by saying that the upper-right and the lower-left rectangles have equal areas—that is, these 2 rectangles with the R's in them have equal areas. That means the area of this one, $a$ times $d$, and the area of this one is $b$ times $c$—so that means $ad$ equals $bc$.

Those 2 results are algebraically equivalent—you can get $ad$ equals $bc$ by cross-multiplying in the first one, in the proportion $a/c = b/d$. Algebraically, they're equivalent, but geometrically they are quite different. One is about proportional sides; the other is about equality between areas. That, again, is a difference in conceptual approach.

There are also Chinese proofs in algebra and number theory. Those proofs proceed by looking for the same pattern in different problems—and then unifying and verifying the procedures at a higher level of abstraction. Proofs like that certainly give us a sense of the "why" behind Chinese problem-solving methods—but again, no axioms and, therefore, no logical deduction from axioms.

Finally, there is a really crucial difference between Chinese geometric proofs and Greek ones. The Chinese do not have indirect proofs in their mathematics. This means that there are types of results that they can't prove—like the irrationality of the square root of 2, for example. That requires an indirect proof.

The Chinese may well have suspected that no rational fraction can represent the square root of 2 with complete accuracy, but they never even say this—let alone try to prove it. This fact about Chinese mathematics—that they don't use proof by contradiction—is of sufficient philosophical interest to have attracted a great deal of scholarly discussion. Let me say something about that.

Why might the Chinese mathematicians not have developed indirect proof? Again, it's hard to know why people don't do something, but let's at least try. First of all, it seems to me, indirect proof is essentially argumentative. We know that some cultures, the Navajo for example, discourage contradiction so as to produce social harmony. Could the Chinese have cared more about social harmony than did the disputatious Greeks?

Again, counterfactual reasoning can be subversive: "Suppose we didn't have an emperor. What then?" So, anyway, these are possible social explanations for the absence of indirect proof in mathematics—although these explanations are really not easy to test.

Some distinguished linguists say that the reason that the Chinese didn't develop indirect proof is that the Chinese language did not support the counterfactual construction "If that were not so, then it would follow that ..." Many linguists hold that language shapes thought.

Of course, once a new concept comes in from somewhere else, the language in question can find ways to describe it—but it's much harder to invent a concept if your language doesn't have words and grammatical constructions suited for that concept. So, I used to believe that, but once I asked a historian of classical China why the Chinese didn't have indirect proofs, and he said: "But in philosophy, they did." He referred me to a paper by Donald Leslie. I'll give you one of Leslie's examples, and you can see what you think about it. This comes from the Confucian philosopher Mencius of the 4th century B.C.E.

Here's Mencius's argument. It's an argument that all people have the same taste for flavors. "Suppose," Mencius says, "that the mouth of each man differed from that of other men in its taste for flavors.... Then why should all men follow Yi Ya in their tastes? But in the matter of tastes everyone models himself on Yi Ya. That is to say, that all the mouths of all men resemble one another." Irrespective of the content of that argument, it does seem to me to be an indirect proof—a proof by contradiction.

So the question isn't why the Chinese didn't have proof by contradiction, but rather why they didn't use them in mathematics. It seems to me—and this is a controversial question, and I'm not an authority on Chinese mathematics—but it seems to me that there are 2 things going on.

One is that Chinese mathematics, although it generalizes and abstracts from the concrete individual case, does not lose sight of that individual case. So it doesn't seem like a scientific way to proceed to say: "Suppose that an angle has a value that we already know that it doesn't have." That's one possibility.

Another thing, in geometry, as we saw when we were studying indirect proofs in Euclid, you remember the proof of Proposition 27, where we said if the alternate interior angles are equal, then the lines are parallel, and I drew a picture where the lines intersected, and when you looked at that you could see, well, those angles really don't look like they're equal anymore? Okay. You can't draw a plausible picture for the counterfactual assumption. So, if you're thinking of actual cases, how can you carry out an argument in geometry if you can't even draw it?

The other thing is that Chinese mathematics really is closer to technology and applied science than it is to the lofty abstractions of the Greeks. There's a rich tradition of observation and a search for discoveries that are useful, not for those that may be impossible.

There is also a sociological explanation as to why mathematics stayed tied to applications and didn't develop independent methods like indirect proof. In classical China, there don't seem to have been many full-time professional mathematicians.

In the Greek-speaking world, in Alexandria especially, there were some people who devoted their lives to doing just mathematics. Sociologists of science argue that professionalization is helpful in a discipline's developing of a sense of autonomy, of having its own methods, and of answering only to its own standards. That didn't really happen in classical China.

All that I've said adds up to just a set of correlations—it's really hard to be sure what causes what. What is clearly true, though, is that classical Chinese society and Greek-speaking society in the 1000 years between Pythagoras and Hypatia of Alexandria—these societies were culturally different, and they produced different, although equally sophisticated, mathematical traditions.

We do know enough to draw some important conclusions, especially since we've looked at 2 separate cases. All right, important conclusion number 1: There is no single straight-line path along which mathematics must develop and progress. In the Middle Ages, the Chinese had made discoveries yet unknown to the West about many types of mathematics, including solving various types of equations—although the Chinese hadn't developed anything like the Greek theory of irrationals.

This same observation—that one culture can know lots more about one type of mathematics than another culture, while the second culture can know more about different types of mathematics than the first—this observation also holds for the classical mathematical traditions of India and Persia, and for the mathematics and mathematical astronomy of the indigenous peoples of the Americas, for example.

Second, looking at mathematics in a different cultural setting—for instance, the preference in Chinese geometry for relationships between areas over proportionality of lines, or the Chinese preference for cut-up-the-area-and-put-it-back-together-again proofs in geometry over the use of counterfactual arguments—indirect proofs—that suggests that there are psychologically different ways of looking at similar mathematical subjects. We already saw this when we looked at the different concepts of space in different languages and different cultures.

Knowing all of this ought to help us do a better job teaching mathematics. Even within the classrooms in one society, we find a variety of different learning styles. None of them is "the best." Some people love algorithms and are really, really good at gaining insight using the arguments of symbolic algebra.

Those people are like Descartes, and Leibniz, and Condorcet. Other people think about problems more visually, and they prefer arguments based on geometric intuition. These people are like Euclid, and Kant, and Yang Hui—the guy who wrote up the Pascal triangle in China and the guy who wrote the *Nine Chapters on the Mathematical Art*. Still, other people love logic above all, and they like to explore counterfactual hypotheses—and those people are like Gauss, and Bolyai, and Lobachevsky, and Socrates.

We should be more aware of these differences as we teach. We should help students build on their individual strengths to help each one learn more effectively. Of course, everybody ought to understand all these different approaches. But everybody in the mathematical community doesn't have to think alike. More progress will be made if multiple perspectives illuminate all the aspects of a given problem. I sound like Ortega y Gasset, don't I?

Finally, I believe that the Chinese example does what I promised when I started this lecture: It gives us another example besides the Western to show us that the relationship between mathematics and culture is reciprocal, is complex, and is fascinating.

In the next lecture, we'll see one more fact about this relationship between mathematics and culture: It can also be adversarial.

# Lecture Thirty-Five
# The Voice of the Critics

**Scope:** There are many people who do not like mathematics: its claims to truth, its claim to have the best method to develop science, and even its nature and subject matter. One piece of evidence is that at every step, the views we have been describing have provoked strong and robust opposition. We will look at 3 classes of critics of mathematics: people who use mathematical and logical methods to attack what they believe are extravagant claims based on other mathematical ideas; critics who grant the value of mathematical methods in their own sphere but who want to set limits beyond which mathematical reasoning should not pass; and critics who regard the whole enterprise as cold, unfeeling, or even totalitarian. Since the critics include some major figures in Western thought, their views should carry considerable weight.

# Outline

**I.** The importance of mathematics in our culture is signaled by the fact that every development we have considered has provoked strong and robust opposition.

   **A.** One class of opponents is not hostile to mathematical methods; rather, they use such methods to attack the "misuse," by others, of mathematical ideas.

   **B.** Another class of opponents also is not hostile to the mathematical method as such but do not want mathematical methods extended beyond their proper domain.

   **C.** The third class of opponents really are opponents: They see mathematical reasoning as cold, unfeeling, imperialistic, destructive of the human spirit, and even totalitarian.

   **D.** We will look at each class of critics in turn.

**II.** The first category typically includes individuals who opposed one view of mathematics while championing another.

   **A.** Aristotle opposed Plato's view of mathematics as concerned with eternally existing objects in favor of seeing it as logical deductions about ideas abstracted from reality.

**B.** Isaac Newton opposed René Descartes' attempt to deduce the laws of nature from self-evident first principles.

    **1.** Newton held that God set up the world according to his own free choice, not mathematical necessity.

    **2.** Nonetheless, Newton's mathematical physics far surpassed anything that had been done before.

**C.** Thomas Robert Malthus's famous *Essay on the Principle of Population* begins with postulates and analyzes the growth of population and of food supply by means of mathematical models, but his chief targets are the predictions by the Marquis de Condorcet and others of continued human progress modeled on that of mathematics and science.

**III.** The second category involves drawing a line that mathematics should not cross.

**A.** Reacting against Cartesian rationalism, Blaise Pascal contrasted the "esprit géometrique" (abstract and precise thought) with what he called the "esprit de finesse" (intuition), holding that each had its own sphere but that "the heart has its reasons which reason does not know."

**B.** In the early 19th century, mathematicians and philosophers like Joseph-Louis Lagrange and Auguste Comte thought that all of science could be reduced to mathematics (and, in Comte's case, that even the study of society could ultimately be reduced to mathematical science).

**C.** But mathematicians like the great Augustin-Louis Cauchy said instead that although we should cultivate the mathematical sciences, they should not be extended beyond their domain.

**D.** As Plato said, mathematics is hypothetical: It involves if-then thinking. Some modern thinkers have called this "instrumental reason" and worried about its amorality.

    **1.** A Renaissance example is Machiavelli's *The Prince* where he tells us how, if you want to rule a state, then you have to do various things, like kill off your enemies or pretend to be religious.

    **2.** Between 1700 and 1810, almost a million slaves were sent to the British West Indies; the fact that the extensive slave system can be documented with an abundance of statistics gathered by private companies and government indicates the instrumental rationality it embodied.

3. The Nazis killed millions of Jews as a textbook exercise in instrumental reasoning, speaking of "the final solution" of the "Jewish problem" and treating the questions of how to transport their victims, how to kill them efficiently, and how to dispose of their bodies as technical questions with technical solutions.

4. As a corrective, computer scientist Joseph Weizenbaum, decrying those who think that human beings are nothing but processors of information, opposed what he called the "imperialism of instrumental reason" and urged that the computer scientist should always remain aware of—and teach—"the limitations of his tools as well as their power."

IV. The third category are those who oppose the method of analysis, the mathematization of nature, and the application of mathematical thought to human affairs, and who grant no saving grace to the subject.

A. Jonathan Swift's satire of statistical thinking, "A Modest Proposal," "solves" the problem of Irish poverty with a computationally sound suggestion that the children involved be eaten.

B. George Berkeley, Bishop of Cloyne, attacked the Newtonian worldview, believing that the mathematically based God of Nature was supplanting the God of Revelation, saying that that Newton's absolute space was nothing but a fiction, and even attacking the calculus as inadequately based in logic.

C. The Romanticism of the 19th century abounds with examples; we give a few.
1. William Wordsworth's poem "The Tables Turned" includes the line "We murder to dissect."
2. The poet Johann Wolfgang von Goethe even opposed Newton's analysis of white light into its component colors.
3. Walt Whitman's poem "When I Heard the Learn'd Astronomer" describes a listener sickened by charts and calculations, driven outside to look up "in perfect silence at the stars."

D. Charles Dickens, in his 1854 novel *Hard Times*, attacks the statistical generalizations about the economy of Victorian Britain that neglect individual misery, and he denounces the analytically based industrial division of labor that neglects the humanity of the workers involved.

**E.** The Russian novelist Yevgeny Zamyatin wrote a novel called *We*, in which mathematical tables of organization are used as instruments of social control.

**F.** Hannah Arendt, in *Eichmann in Jerusalem*, said that excusing one's actions because one is a "mere functionary" is like a criminal saying he did only what was statistically expected.

**G.** Fyodor Dostoyevsky, in his *Notes from Underground*, argues that no matter what is in a person's rational self-interest, no matter what natural law dictates, one should choose one's wildest whim "to show he's a man and not a piano key."

**V.** The power, certainty, and authority of mathematics can be used in a variety of ways.

**A.** Mathematics can be an ally of liberalism, as we have seen in the cases of Voltaire, Thomas Jefferson, and the Marquis de Condorcet.

**B.** It can be used as a way of establishing an unchallengeably authoritarian government, as Plato wanted to use it.

**C.** It can be used to avoid hard thinking by treating a problem as solved once we have a set of numbers, as Stephen Jay Gould pointed out.

**D.** It can be used as a tool, and the users can get so wrapped up in the problems involved that they lose sight of the ethical and human context.

**E.** It can be a tool, used critically, to evaluate those things capable of precise understanding and to recognize the limits of such precision; this is the point of view we have tried to champion in this course.

### Essential Reading:

Dickens, *Hard Times*.

Grabiner, "The Centrality of Mathematics."

Wordsworth, "The Tables Turned."

### Suggested Reading:

Arendt, *Eichmann in Jerusalem*.

Bayley, "*Hard Times* and Statistics."

Dostoyevsky, *Notes from Underground*.

Malthus, *An Essay on the Principle of Population.*

Pascal, *Pensées.*

Weizenbaum, "Against the Imperialism of Instrumental Reason."

Zamyatin, *We.*

**Questions to Consider:**

1. What explains the disdain so many people have for the mathematical way of thinking? To what extent is this disdain justified?

2. If if-then thinking truly does not inquire into the moral or social meaning of either the premises or the reasons, how can it be taught in a socially responsible manner?

# Lecture Thirty-Five—Transcript
## The Voice of the Critics

Welcome back. Today, we're going to hear from the other side—and there is another side. In fact, the importance of mathematics to Western culture is signaled by the fact that every development we have considered so far has provoked strong and robust opposition. That's worth repeating. The importance of mathematics to our culture is signaled by the fact that every development we've considered has provoked strong and robust opposition.

It's easier to get a handle on this opposition if we observe that the critics of mathematics fall basically into 3 different groups. First: Those who use mathematical methods themselves, but who attack their misuse by others. Second: Those who, although not hostile to mathematical methods, do not want mathematical methods extended beyond what these people see as the proper domain for such methods. Third: Those who see mathematical reasoning as cold, unfeeling, imperialistic, destructive of the human spirit, reducing people to statistics, even totalitarian.

In this lecture, I'll look at all 3 classes of critics. But I'll spend most of our time together on the third group. I want to focus on the third group because if there's one thing we should have learned from all this philosophy—especially from the power of the reasoning that starts with "if this is not the case"—it ought to be clear that we have a lot to learn from our intellectual opponents.

So, let's start with the first category—those critics who have opposed one point of view about mathematics, while championing another point of view about mathematics. Actually, a great deal of the revolutionary progress in mathematics itself is like that. For instance, the Greeks learned a lot about geometry from the Egyptians and the Babylonians, but the Greeks didn't accept the philosophy of mathematics of either of those cultures. The Greeks instead insisted that the merely plausible, or the visually apparent, or the closely approximate—those weren't good enough: Geometrical truths needed to be proved logically from self-evident assumptions.

Another example: Descartes thought that algebra was too limited to the formulaic statements of rules, and that geometry was too dependent on visual representations. So, he invented analytic geometry, which he said took the best traits of geometry and algebra and "corrected the faults of one by the other." Likewise, Gauss, Lobachevsky, Bolyai, and Riemann accepted the need for a logical structure for geometry, but they changed utterly the foundation on which geometry rested. They changed it from resting on one set of so-called self-evident necessary truths to resting on any one of many interesting sets of assumptions that are ultimately the result of human choice.

Philosophers, also, have championed one view of mathematics while accusing others of getting it all wrong. For instance, Aristotle opposed Plato's theory of Forms. Since it is mathematics that Plato used to make plausible the view that there is an intelligible world of non-material objects, and since Aristotle rejected that transcendental world, Aristotle attacked Plato's view that the objects of mathematics exist in such a transcendental realm. Instead, Aristotle treats mathematical objects as abstractions from the world of sense experience. Aristotle isn't against mathematics—after all, he's the great prophet of demonstrative science, but not mathematics described Plato's way.

Also, as we've seen, Newton opposed Descartes's attempt to deduce the laws of nature from self-evident first principles. Newton thought that God set up the world by free choice—not mathematical necessity. Nevertheless, Newton created the most mathematically sophisticated science that the world had yet known. You'll recall how Laplace's mathematically based determinism was attacked by Maxwell's argument for free will based on mathematically sophisticated statistics.

My last example of this type is new in this lecture, and it comes from the work of Malthus. Malthus is best known for his *Essay on Population* of 1798. But you need to hear the full title: *An Essay on the Principle of Population, as it Affects the Future Improvement of Society with Remarks on the Speculation of Mr. Godwin, M. Condorcet, and Other Writers.* Monsieur Condorcet we have already met. As for Godwin, William Godwin was a political philosopher who wrote a book called *Enquiry Concerning Political Justice.* In that book, Godwin said that humanity would ultimately reach a utopian state, and that the human population would never surpass the means of subsistence. The operations of the intellect, Godwin thought, would produce an ideal kind of existence.

Condorcet, besides as we've already seen promoting the idea of universal human progress based on the progress of science, Condorcet also addressed the problem of population. Condorcet said that people, of course, would limit population should that become necessary; in the future they would be rid of the "absurd prejudices of superstition" that might prevent them from doing so. Predictions of human progress like Condorcet's and Godwin's owe a lot to science, and reason, and mathematics.

But Malthus was less optimistic than Godwin and Condorcet, and Malthus used mathematical methods to refute his opponents. Here's how he starts his argument: "I think I may fairly make 2 postulata." That's 2 postulates, and what are Malthus's postulates?

First, that food is necessary to the existence of man. Second, that the passion between the sexes is necessary and will remain nearly in its present state. Once you grant his self-evident assumptions, he says, that it follows that "the power of population is indefinitely greater than the power in the earth to produce subsistence for man."

Famously, Malthus said that population increases in geometrical ratio, while subsistence increases only in an arithmetical ratio. What that means is that every year the population is multiplied by some number; every year the food supply just has a number added to it. Eventually, multiplication will always outstrip addition. To show this, Malthus gives a series of numbers; they aren't empirical numbers. Let's say these are the numbers, but it's even clearer when we illustrate it with a graph. So, look at the graph. What happens when the population reaches this point, where it's about to outstrip the food supply? Then the population can't go on increasing, because there's not enough food—so the result is, Malthus says: "misery or vice." So, in opposition to the use of mathematics and science by Condorcet and others to argue for perpetual human progress, Malthus uses the postulate-and-consequences approach and mathematical models to argue that this progress will not occur.

Malthus's arguments continue to be influential, informing debates about population control in our own time. For our present purposes, though, the key point is that Malthus is one more person who uses mathematical methods to shoot down arguments made by others who draw what he sees as a wrong conclusion from mathematics.

Let's go on to my second category: people who grant that mathematics is important and valuable, but who still hold that there are areas in which mathematics should not go. We begin with Pascal. Pascal opposed what he saw as the over-rationalism of Descartes. Pascal contrasted what he called the *"esprit géometrique"*—that is, "abstract and precise thought"—with what he called the *"esprit de finesse"*—that is, "intuitive understanding." Pascal thought that each of these approaches had its own sphere of appropriate action, but—and this may be the most famous quotation from Pascal ever—"the heart has its reasons which reason does not know."

I'm going to introduce another critic, a critic who arose to attack later people like Lagrange, Laplace, Francis Hutcheson, and their followers.

First, let me show you in some more detail what there was to attack. Lagrange, whom we've already met, believed that all of physics could be reduced to mathematics. One of Lagrange's followers was the philosopher Auguste Comte.

Comte pushed Lagrange's idea of reducing physics to mathematics much farther. As physics can be reduced to mathematics, Comte said, so chemistry can be reduced to physics, so biology can be reduced to chemistry, so psychology can be reduced to biology, and a new subject, sociology—Comte invented the name, by the way—sociology could be reduced to psychology. So ultimately, everything rests on mathematics.

Many other scientists of the 18th and 19th centuries believed similar things. For instance, Laplace and others, they thought that mathematical models could perform better than human judges in the courtroom. Francis Hutcheson (whom we've already met) and later the utilitarians thought that you could choose the best action by determining, maybe by using calculus, which action produced the greatest good for the greatest number.

Now for the critic: The leading French mathematician in the first half of the 19th century was Augustin-Louis Cauchy. Cauchy thought that all of those ideas crossed a boundary that mathematics shouldn't cross. Here's what Cauchy wrote in 1821: "Indeed, let us assiduously cultivate the mathematical sciences, but," said Cauchy, "let us not imagine that one can attack history with formulas, or give for sanction to morality theorems of algebra or integral calculus."

Other critics zeroed in on the lack of ethical concern that is involved in pure "if-then" thinking—a mode of thought so common in Greek geometry. A Renaissance instance of pure "if-then" thinking, although not mathematical in content, is what is often called the first work of modern political science—namely *The Prince* by Machiavelli. Machiavelli explained that if you want to be a successful ruler, then a scientific study of historical examples shows that you need to do particular things: like kill off your enemies, or pretend to be religious. I mention Machiavelli because he's often considered the master of using reason as the way to get from the "if" to the "then," from the postulated goal to the unfortunately necessary means. He's been criticized so much for this that people often accuse somebody who believes that the ends justify the means of being "Machiavellian." Reason, say these critics, should be more than just an instrument. We should use logic, reason, mathematical models, and the like to examine the consequences of our actions—not just to do a job set for us by somebody else.

Mathematics, say these critics, is being improperly used if you use it to work out the "then" part of something where the "if" part is morally wrong. These critics tell us: Don't fall in love with reason as an instrument. Pay attention to what you're doing. Here's one of these critics, the historian of slavery Robin Blackburn. Blackburn first tells us that between 1700 and 1810, about a million slaves were sent from Africa to the British West Indies. Blackburn then says the fact that the extensive slave trade can be documented with an abundance of statistics, gathered by private companies and government agencies alike, that fact "indicates the instrumental rationality that it embodied." Blackburn doesn't think there's anything wrong with using statistics, but, hey, think about what you're counting. Shouldn't you see the people involved?

The Nazis killed millions of Jews, and Slavs, and Gypsies, and homosexuals, and the mentally ill—sometimes speaking about doing this as though it were a textbook exercise in instrumental reason. They spoke of the "final solution" of the "Jewish problem." The Nazis treated the questions of how to transport their victims, how to kill them efficiently, and how to dispose of their bodies as technical questions with technical solutions.

One more example: Computer scientists use mathematics to understand how to process information. Some computer scientists have claimed that all thought, including human thought about the arts and human emotions, all thought is nothing more than information processing. In opposition to these views, the late computer scientist Joseph Weizenbaum denounced what he called "the imperialism of instrumental reason." Weizenbaum despaired of those who thought that human beings are, like computers, nothing more than processors of information—nothing more than mathematicizing machines. Weizenbaum urged that the computer scientist should always remain aware of—and teach—the limitations of his tools as well as their power.

Having heard that eloquent statement by one of the second category of critics, let us now move on to the third category: those who oppose the method of analysis, the mathematization of nature, and any application of mathematical or statistical thought to control human affairs.

In the 18th century, a striking example comes from Jonathan Swift. I'm not talking about Swift's most famous book, *Gulliver's Travels*—although if you read it, Gulliver's third voyage, to the Grand Academy of Lagado, is in good part a satire attacking the scientists of the Royal Society of London. But I want to focus on Swift's *Modest Proposal* of 1729. Swift's full title was *A Modest Proposal for Preventing the Children of Poor People in Ireland Being a Burden to Their Parents or Country, and for Making Them Beneficial to the Public*. Swift satirically proposed that Ireland's poor could improve their economic lot by selling their children as food for the rich.

What Swift was satirizing here was the statistical study of society, which was then called "political arithmetic," where numbers and averages were used to justify governmental policies—both at home and abroad, regardless of the effects on individuals. But Swift, unlike our earlier categories of critics, didn't want to reform political arithmetic; he wanted to discredit it. If you read Swift's *Modest Proposal*, you'll see that he did a pretty good job.

Also in the 18th century was somebody else we've met, Bishop George Berkeley. I briefly mentioned Berkeley earlier for his opposition to Newton's idea of absolute space—and Berkeley's emphasis, instead, on the way people actually perceive space; we see angles and not distances. That was just one instance, though, of Berkeley's opposition to Newton's whole world view. Why did he oppose it? One thing is that Berkeley didn't like the way belief in the mathematically based God of Nature of the Newtonians was supplanting the God of Revelation. For religious and philosophical reasons, Berkeley not only attacked Newton's absolute space as a fiction, Berkeley even attacked the calculus as inadequately based on logic. By the way, Berkeley's criticisms of the calculus of his day had merit. Mathematicians in the 18th century saw Berkeley as a formidable opponent, and that's what Berkeley wanted to be.

In the 19th century, the philosophy of Romanticism abounds with examples of distaste for reason, distaste for analysis, and distaste for mathematics. For instance, the English poet William Wordsworth wrote a poem called "The Tables Turned," and in that poem he attacked the method of analysis.

Wordsworth wrote:

> Sweet is the lore which Nature brings;
> Our meddling intellect
> Mis-shapes the beauteous forms of things:—
> We murder to dissect.

Notice that it's the analytical mind—not a physical scalpel—that Wordsworth is criticizing.

The German poet Johann Wolfgang von Goethe opposed the analysis of white light into colors that Newton had carried out, preferring to describe the feelings evoked by color.

The American Romantic poet Walt Whitman wrote a poem called "When I Heard the Learn'd Astronomer." The poem is about somebody who goes to a lecture on astronomy. The astronomer did his calculations, and he drew his charts. Then, Whitman has the listener say that he became tired and sick, and he leaves the lecture room and runs outside, where he "looked up in perfect silence at the stars."

Quite a distinguished list of writers that we're looking at here. Next is Charles Dickens. Dickens attacks the statistical view of humans in society in his 1854 novel, *Hard Times*. Dickens has one character, a workman, eloquently attack the analytically based idea of the division of labor. The character says that this treats workers as though they were nothing more than "figures in a sum."

Another character in Dickens's novel, *Hard Times*, is called Thomas Gradgrind. Gradgrind heads a school in which nobody is allowed to read fiction. Facts are what matter to Gradgrind—nobody is allowed to read fiction, and the students are addressed by number, if you want to ask a question: "Girl number twenty," like that. The teacher in the school—if you know Dickens, he gives people really good names—the teacher is called Mr. M'Choakumchild. The teacher asks this girl—we saw this already in Lecture Two—Don't you think that Britain is prosperous because there is so much wealth (such a high average annual income, we might say). The girl replied: "[I can't tell] unless I [know] who had got the money, and whether any of it was mine." When the teacher says that in a city of a million inhabitants, only 25 starve to death in the streets in the course of a year, and the girl says: "It must be just as hard [upon] those who were starved, whether the others were a million, or a million million."

Later on in the novel, Gradgrind's daughter comes to see him for advice. She is worried about whether to marry a certain man. The reason she's worried is, she says: "Life is short." Here's her statistical father's reply: "It is short, no doubt, my dear. Still, the average duration of human life is proved to have increased of late years. The calculations of various life assurance and annuity offices ... have established this fact." "I speak of my own life, father," says the girl. "O Indeed? Still, I need not point out to you, Louisa, that it is governed by the laws which govern lives in the aggregate." The father is a lot of help.

Near the end of the book, Gradgrind's son robs his brother-in-law's bank. Gradgrind is understandably upset. But his son says: "I don't see why you should be shocked father. Given a population of a certain size, a certain percentage will be dishonest," said the son, "I've heard you talk a hundred times of its being the law."

Dickens just hates all of this—whether it comes from statisticians like Quetelet or from those who worship Adam Smith and his analysis of the laissez-faire marketplace. In one well-crafted sentence, Dickens tells us where his loyalties lie—not with the individual maximizing his own gain and hoping that this will promote the welfare of society. "The Good Samaritan," Dickens says, "was a bad economist."

The Russian novelist Yevgeny Zamyatin wrote an anti-utopian novel called *We*. If you've read *1984* by George Orwell, you'll be interested to know that Zamyatin's novel was one of Orwell's major influences. In the novel *We*, mathematical tables of organization are used as instruments of social control. The fictional society thinks of itself as "the greatest and most rational civilization in history."

People are called "numbers," rather than "people" or "citizens," and that's also all the name they have. The book's hero is called D-503. He has a girlfriend, who is called I-330. He can see her when her name is entered in his "Sexual Table"—see tables, numbers, are the organizing principle of the whole society. He can see her when her name is entered in his Sexual Table, but she has to bring a pink coupon with her for him to collect.

The novel portrays these 2 rebelling against the established order—against this mathematical tyranny. But the society calls these rebels "a group of numbers who have betrayed Reason." At the end of the novel, D-503 undergoes an operation to remove his imagination, and the novel ends with these words: "Reason must prevail."

These novels are scary enough, but the real world can be worse. Hannah Arendt wrote a book called *Eichmann in Jerusalem*. It's about the trial of the man who was the chief organizer of the murder of millions of Jews in Nazi Germany. Eichmann portrayed himself as carrying out his duties, "just following orders."

Here's what Hannah Arendt says about Eichmann's defense:

> If the defendant excuses himself on the ground that he acted not as a man but as a mere functionary whose functions could just as easily have been carried out by anyone else, it is as if a criminal

pointed to the statistics on crime—which set forth that so-and-so many crimes per day are committed in such-and-such a place—and declared that he only did what was statistically expected, that it was mere accident that he did it and not somebody else, since after all somebody had to do it.

That probably reminds you of Quetelet and Dickens. But we're going to get Zamyatin and Orwell too, in a minute. You don't have to buy Hannah Arendt's total explanation of the Holocaust to see the point of what she's saying here. She goes on:

The essence of totalitarian government, and perhaps the nature of every bureaucracy, is to make functionaries and mere cogs in the administrative machinery out of men, and thus to dehumanize them.

As we reflect on these examples, let's again go back to the 19th century and allow the Russian novelist Fyodor Dostoyevsky to drive all these points home. Dostoyevsky wrote a book called *Notes from Underground*. In that book, his main character maintains that human beings aren't in this world to follow the laws of reason.

What [man] wants to preserve [Dostoyevsky's character says] is precisely his noxious fancies and vulgar trivialities, if only to assure himself that men are still men … and not piano keys simply responding to the laws of nature…. Even if man were nothing but a piano key [says Dostoyevsky] even if this could be demonstrated to him mathematically—even then, he … would pull some trick out of sheer ingratitude, just to make his point….

Dostoyevsky goes on:

You may say that this too can be calculated in advance and entered on the timetable—chaos, swearing, and all—and that the very possibility of such a calculation would prevent it, so that sanity would prevail. Oh no! In that case man would go insane on purpose, just to be immune from reason…. It seems to me [Dostoyevsky's narrator concludes] that the meaning of man's life consists in proving to himself every minute that he's a man and not a piano key. [Or, as Joseph Weizenbaum would say, a human being and not merely a processor of symbolic information.]

Let's see if I can draw a conclusion from this lecture. We've looked—throughout the course—at the power, certainty, and authority of mathematics. These things can be used in a variety of ways. Mathematics can be an ally of classical liberalism—as we've seen it used by Voltaire, by Jefferson, and by Condorcet.

It can be used as a way of establishing an unchallengeably authoritarian government—as Plato wanted to use it, and as Zamyatin showed it being used. Mathematics can be used to avoid hard thinking by treating a problem as solved once we have a set of numbers—as Dickens's Mr. Gradgrind tried to do, and as Stephen Jay Gould warned us not to do.

Mathematics can be used as an instrument, as a tool, and the people who use the tool can get so wrapped up in the technical problems involved that they lose sight of the ethical and human consequences.

Finally, though, mathematics can be a tool used critically to evaluate those things capable of precise understanding, and also to point out the limits of that kind of understanding. Philosophers, scientists, mathematicians, and members of society at large—all need mathematics to do these things: to evaluate those things capable of precise understanding, and also to recognize the limits of that kind of understanding. This is the point of view I've tried to advocate during this course: to balance probability and certainty, to learn both the value of precision and the existence of uncertainty and error. As we approach the concluding lecture, let's keep in mind both the strengths and the limitations of mathematics.

# Lecture Thirty-Six
## Mathematics and the Modern World

**Scope:**  We return to the claims we made at the beginning: Probability and statistics have greatly changed the way people think. We review the principal points we have made. We then turn to 2 paradoxical conclusions from the story we have been telling. First: Even matters of truth and certainty and knowledge are conditioned by culture as well as by necessity. Second: Randomness and chance can be intellectually tamed by the rigorous reasoning of mathematics; in fact, probability theory (like geometry) can be based on axioms. But since mathematics has continued to grow and its interactions with the rest of thought and society have also grown, we briefly introduce 4 of the most striking modern philosophical questions that we have not yet addressed: entropy and why time does not run backward; chaos theory; the fact that we can prove that we cannot prove the consistency of mathematics; and the many questions, from the computer's role in art to whether computers possess actual intelligence, raised by the computer revolution. We close with the hope that the historical story we have told in these lectures provides us with a level of understanding equal to the challenges of these modern developments—and whatever the future may bring.

## Outline

I.   We began this course by saying that probability, statistics, and geometry have changed the way people think about the world.

    **A.**  We have shown how pervasive statistical thinking is in modern society.

    **B.**  We have shown how the use of probability and statistics helped give birth to the social sciences, affected debates about free will versus determinism, and produced new approaches to the natural sciences that put the very notion of causality into question.

    **C.**  We have shown how the ideal of certainty, present in geometry since the time of Plato and Aristotle, shaped ideas of certainty and truth in philosophy, and how this same ideal became a model for political and ethical thought.

**D.** Through our analysis of logical argument and statistical reasoning, we have provided tools for evaluating arguments, whether these are quantitative or qualitative, on the basis of good statistical practice, knowledge of how to calculate probabilities, and analysis of arguments for their validity and soundness.

**II.** We have seen also that the history of mathematics is not just a triumphal story of greater knowledge and precision.

    **A.** Culture, including philosophy, art, literature, and social arrangements, has affected the questions that mathematicians ask and the methods that they use.

    **B.** And we have seen that mathematics has its critics and opponents, often for good reason.

**III.** We have by no means exhausted the ways mathematics has affected human thought, especially philosophy and science; we should at least mention some of the outstanding modern interactions between mathematics and philosophy, of which 4 seem especially important.

    **A.** First, the statistical view of the universe is closely related to the 19th-century concept of entropy.

        **1.** Entropy is a measure of the disorder of a system.

        **2.** For instance, a room that is half full of cold air, half of hot, is highly ordered and has relatively low entropy.

        **3.** The kinetic theory of gases helps us see how in such a room, the particles will mix, and the room will approach a state in which it is uniformly warm.

        **4.** The uniformly warm room stays that way; the slow particles do not go off to half the room and the fast ones to the other, thus dividing cold and hot again and going from disorder to order. This, in short, is the second law of thermodynamics.

        **5.** That the entropy in the universe increases over time implies that time's arrow always goes in one direction.

        **6.** From philosophers who imagined that ultimately the universe would all be the same temperature and all activity would cease, to those who used the concept of naturally increasing disorder to describe society or to claim that evolution by natural selection could not occur, many thinkers have drawn far-ranging conclusions from this particular contribution to the statistical study of particles in motion.

**B.** Second, the limitations of the Laplacian model of deterministic science are addressed by what is called chaos theory.

    **1.** The key idea here is that a very small change in the initial conditions of a system can produce very large changes in the eventual outcome.

    **2.** One metaphor often used is that the fluttering of the wings of a butterfly in one part of the world can produce a major storm somewhere else.

    **3.** So Laplace's infinite intelligence has its work cut out for it; prediction, even in principle, is much harder than anybody had imagined.

    **4.** The computer models that have produced these conclusions have also produced amazing visual objects, like fractals, and the use of these in art has further transformed the modern imagination.

**C.** Third, returning to the idea of proof and certainty, the mathematician David Hilbert in 1900 proposed that the consistency of mathematics should—and could—be proved.

    **1.** Hilbert hoped that, just as we can prove that certain positions of the pieces cannot occur in a game like chess because of the nature of its rules, we could prove that no formula like "A and not-A" could occur in mathematics.

    **2.** So mathematics would continue to progress, and we could eventually prove every true result in our axiomatized system, as René Descartes had dreamed.

    **3.** However, Kurt Gödel proved that in any mathematical system rich enough to allow the known theory of numbers, there are propositions whose truth or falsity cannot be proved within the system; one such proposition is the one that asserts the consistency of the system.

    **4.** As a result, some say that there is no certainty anywhere—not even in mathematics.

**D.** Finally, a combination of mathematics, logic, and the divide-and-conquer approach to problems produced a revolutionary invention: the digital computer.

    **1.** Computers have transformed the economy, communication, and education in the modern world.

    **2.** The properties of computers raise many philosophical questions, notably, can computers think? Could they develop consciousness?

**IV.** We might also mention the idea of the infinite.

    **A.** Infinity has been an important idea for theology.

    **B.** The impact of believing in an infinite universe has been great.

    **C.** Still, until recently with the work of Georg Cantor, the idea of the infinite really belongs more to philosophy than to mathematics.

**V.** We look back on the whole course.

    **A.** With the increasing role of mathematics in all aspects of society in the modern world, it is more important than ever to understand how mathematics interacts with all areas of human thought.

    **B.** In this course, I have tried to provide a historical perspective on all of this.

    **C.** Even given the items listed in the present lecture, I think that my initial point is correct: Probability and statistics, and Euclidean (and non-Euclidean) geometry, are the branches of mathematics that have had, by far, the greatest impact on Western thought.

    **D.** My hope is that this course has enhanced your ability to get a handle on these crucially important questions and also inspired you to pursue the topics further.

**Essential Reading:**

Davis and Hersh, *The Mathematical Experience*.

**Suggested Reading:**

Blum, *Time's Arrow and Evolution*.

Brown, *Philosophy of Mathematics*.

Gamow, *One, Two, Three—Infinity*.

Gleick, *Chaos*.

Jammer, "Entropy."

Koyré, *From the Closed World to the Infinite Universe*.

Newman and Nagel, *Gödel's Proof*.

Peter, *Playing with Infinity*.

Stoppard, *Arcadia*.

Von Baeyer, *Maxwell's Demon*.

**Questions to Consider:**

1. How many ways have you interacted with computers during the past week? How would your parents, when they were your age, have answered this question? Your grandparents? (The fact that they might never have heard of computers is relevant here.) Do these different answers signal a revolution in the human experience?

2. Do the mathematically based conclusions described in this lecture—that the "film" of the history of the universe cannot be run backward, or that the consistency of our mathematics cannot be proved—seem to you to challenge our intuition or even our sense of the stability of the world? Since these conclusions come, respectively, from probability and statistics and from the tradition of proof and certainty, can you fit them into the framework provided by these disciplines?

# Lecture Thirty-Six—Transcript
# Mathematics and the Modern World

Hello. I started this course by saying that probability, statistics, and geometry have changed the way people think about the world. In this, our last lecture together, I'll begin with a brief review of where we've been. But there's also a lot that I haven't even touched on. So, I want to use the greater part of this concluding lecture to describe some important recent interactions between mathematics and philosophy. Finally, at the end, I owe you a summing-up.

Earlier in the course, then, we saw how pervasive statistical thinking is in modern society. We watched probability and statistics give birth to the social sciences, affect debates about free will and determinism, and produce new approaches to the natural sciences that challenged the very idea of cause and effect.

From Stephen Jay Gould to Pascal, from Laplace to Quetelet to Maxwell, from Niels Bohr to Albert Einstein, these topics have involved us with some of the major figures in the history of science. We've also seen how the ideal of certainty, present in geometry at least since the time of Plato and Aristotle, helped shape the ideas of certainty and truth in philosophy. This same ideal became a model for political and ethical thought.

From Euclid to Lobachevsky, we've seen the power of reasoning as applied within mathematics itself. Through thinkers as different as Spinoza and Jefferson, as Descartes, and Newton, and Malthus, we've seen the model of demonstrative science as an actor in philosophical, social, and political debate as well as scientific. We've described, and practiced using, the most historically influential tools for evaluating arguments—both quantitative arguments and qualitative arguments. For quantitative arguments, we've used the tools of good statistical practice and knowing how to calculate probabilities.

On the qualitative side, we've tested arguments for validity. All these methods, derived from mathematics, are crucial as we live in a world filled with clashing views and with oceans of quantitative data.

That's not all. We've seen how uncertainty and randomness have been "tamed" by mathematical treatment—both in probability theory as a mathematical topic and in its application to the sciences and to society. In fact, in the $20^{th}$ century, probability theory itself was given an abstract axiomatic foundation. Euclid would be proud, and so would all of those philosophers—like Plato, Voltaire, Kant, and Condorcet, who championed the ability of mathematics to order the apparent disorder of the world.

Finally, we've seen that the history of mathematics is not just a story about how mathematicians, oblivious to the rest of culture and life, progress in a straight line toward greater knowledge and precision. Culture—including philosophy, art, literature, and social arrangements— affects the questions mathematicians ask and the methods that they use— whether the mathematicians are in ancient Greece, the Islamic world, classical China, or modern Europe and America. We've seen that mathematics and its uses have critics and opponents—sometimes for very good reasons. Still, I've by no means exhausted the ways mathematics has affected human thought, especially how it has affected philosophy and science. So, in this last lecture, I want at least to mention some of the most important modern interactions between mathematics and philosophy.

I'm going beyond the examples of the course—going beyond geometry, probability, and statistics that we've emphasized. But I think that the glimpses beyond will reinforce many of my generalizations, and I hope they'll be rewarding and interesting in their own right.

I think that 4 of these modern interactions between mathematics and philosophy are especially important. These concern entropy, chaos theory, proofs about proof, and computing. First, entropy: Entropy is a fancy word as a measure of the disorder of a system. This is best illustrated with examples. For instance, consider a room that is half full of cold air and half full of hot air—that is, highly ordered. It's said, therefore, to have low entropy. Looking at this picture, suppose the blue particles are slow-moving, so the air in their half of the room is cold; and the yellow ones are fast-moving, so the air in their half of the room is hot. Think about what's going to happen to the air particles in this room as they mix. The particles move around, and mix, and bang into each other and transfer energy to each other. Pretty soon the speeds of the particles are distributed in the same way all over the room. So, the room approaches a state in which it is no longer half hot and half cold, but uniformly warm—with fast and slow particles mixed all over. Now it is said to have higher entropy.

Here's the important point—the uniformly warm room stays that way. The slow particles do not all go back into one half of the room again, while the fast ones migrate back into the other half—because if that happened, the situation in the room would go, with no outside intervention, from disorder to order. That doesn't seem to happen. The entropy, or disorder, in any closed system—including in the whole universe—tends to increase over time. That entropy increases with time is the essential point of the Second Law of Thermodynamics. The First Law of Thermodynamics is the conservation of energy. It's a statistical law, the Second Law of Thermodynamics. It isn't impossible for the cold and hot particles to separate all by themselves; it's just very, very, very, very improbable.

Or, as our friend James Clerk Maxwell once put it: "The Second Law of Thermodynamics has the same degree of truth as the statement that if you throw a tumblerful of water into the sea, you cannot get the same tumblerful of water out again."

"Entropy always increases" is a statistical statement, but experience certainly bears it out—and the philosophical implications have been immense. First, physics now seems to be telling us, with this idea of entropy, that time cannot be reversed. See, if you had a movie about a billiard ball banging around on a billiard table, assuming that no energy was lost to friction, you could run the movie backwards, and you couldn't tell the difference.

But remember how, back in Lecture Ten, I squirted perfume into the room from an atomizer and later I said that the perfume smell was all over the room? If we ran that movie backwards, you could tell it was backwards. Another common example: You can scramble an egg, but you can't unscramble it again. Time's arrow runs only one way.

Maxwell had an interesting question about the situation where the warm particles and the cold particles were mixed: "Could we reverse the process"? Maxwell thought about this question a lot, about whether we could intervene in this statistical situation to get the slow-moving (or cold) particles all back together by themselves again. Maxwell imagined a little, tiny being—known in the literature as "Maxwell's demon"—who opens a door to let only the slow particles (the cold ones) into half of the room, and only the fast particles (the hot ones) into the other half. You might want to follow the arrows in these schematic diagrams, and you can see this process diagrammed. But there doesn't seem to be any physical mechanism that can make this happen. So, entropy tends to increase—Second Law of Thermodynamics.

Some thinkers—beginning with the physicist Lord Kelvin and another person whom we've already met, Hermann Helmholtz—have foretold a "heat death" for the universe. That heat death would happen when eventually everything is moving at the same speed, the same temperature, so there can't be any change any more. The American thinker Henry Adams applied the concept of inevitably increasing entropy to history and social organization—as revealed in the title of his book, *The Degradation of the Democratic Dogma*.

I hope those examples indicate that "entropy" is an idea worth learning more about. The road to understanding entropy lies in something you already know a lot about—the statistical view of physical nature developed in the 19$^{th}$ century.

Now let us turn to what is called "chaos theory." I'm going to start with a simple example that was already discussed by Laplace. Flip a coin that lands on the floor. Is it going to be heads or tails? Let me do it. We treat the outcome as random because we don't know what will happen, but, in fact, the outcome is completely determined—determined by the shape and mass distribution in the coin, by the angle at which we flip it, by the force of the thumb that does the flipping, by the temperature of the air in the room, the elasticity of the covering of the floor, and so on.

If we change any of these conditions even slightly, the outcome can change substantially. I mean, suppose heads you win a million dollars; tails you lose a million dollars. That's a big difference. Of course, it's all completely determined and, in principle, predictable—but we can't predict it at all; very tiny change in the initial conditions can make a really large change in your life as you bet.

There are lots of things like that coin example. There is a subject in mathematics called "dynamical systems" that studies mathematical models of many kinds of systems that undergo change. Some changing systems have this special property: Their behavior changes substantially with only very small changes in the initial conditions.

A system that behaves that way is called a "chaotic system." The earth's weather is probably the most common real-world example of a chaotic system. We're really good at predicting weather in the very short run, but we're very bad at predicting it over longer time horizons—and that's because small changes can have huge repercussions.

There's a lovely metaphor that illustrates how small changes in initial conditions produce major changes farther down the line: Does the flapping of a butterfly's wings in Brazil cause a tornado in Texas? That question was the title of a lecture given in 1972 by Ed Lorenz of MIT. Lorenz is the father of modern chaos theory. (If you're into science fiction, by the way, you may recognize that "one butterfly can change the world" idea as the premise of a story by Ray Bradbury about time travel called "The Sound of Thunder.") The calculation of the long-term results for chaotic systems is complex, and only computers can do them.

Because a tiny change produces changes later on that get greater and greater over time, the behavior of the system looks as though it's random—hence the term "chaos" or "chaotic system." This name is used even though the behavior of the system is completely defined by the laws of physics and the initial conditions, like the coin flip, with no actual random elements involved. Some people like to call this "deterministic chaos." But Laplace's infinite intelligence has its work cut out for it. Prediction of what will happen in the future, even when we know the laws, is much harder than anybody had imagined.

The computer models and mathematical systems that have produced these conclusions have also produced amazing pictures of them. These include the visual objects called "fractals." A "fractal" is a geometric shape that can be split into parts, where each part is a reduced-size copy of the whole shape. Fractals are very sensitive to small changes in initial conditions. Here's a picture of a fractal.

Pictures of fractals have taken on a cultural significance of their own. The use of them in art has further transformed the modern creative imagination. The one I'm showing you is the most famous—it's called "the Mandelbrot set," named after the father of the modern study of fractals, Benoit Mandelbrot. But fractal geometry occurs in many other places—from African hair styles to the growth of plants.

If you find this material about fractals and chaos especially interesting, I recommend the book in the readings by James Gleick. Also, Stephen Strogatz has a Teaching Company course devoted to this topic.

Tom Stoppard—the playwright, the guy who wrote *Rosencrantz and Guildenstern Are Dead*—addresses the Second Law of Thermodynamics, chaos, and fractals in his recent play, *Arcadia*. Stoppard uses all of those ideas as metaphors for the breakdown of the classical ordered world.

For my third example today, let's return to pure mathematics, to the ideas of proof and certainty. But I think this one is going to be just as exciting for you as its predecessors. Here's the question: Can we prove that mathematics is consistent? We've seen in this course how far mathematics has gone beyond the self-evident. The invention of non-Euclidean geometry and non-commutative algebras (that's where a times b is not b times a), and also the theory of infinite sets—these things made mathematicians wonder: "How much of what we think is true will turn out not to be?" or: "How do we distinguish between what is necessarily true and what is merely what we're used to seeing?" In 1900, the great German mathematician David Hilbert got a very ambitious idea. We can prove a lot of things, Hilbert said, so maybe we can prove the logical consistency of mathematics itself. That would be really wonderful: a proof about all possible proofs within mathematics. How can we do that?

To find such a proof, Hilbert said, we need to think of mathematics as a formal system, where every concept is expressed by unambiguous symbols, and all the valid reasoning about mathematics is expressed by explicit formal rules. We do that with elementary algebra, so that gives us some idea of how it could be done. Once we have this formal model of mathematics—that is, everything reduced to unambiguous symbols obeying explicit formal rules—remember Leibniz wanted to do that, everything reduced to unambiguous symbols obeying explicit formal rules, everything in mathematics—Hilbert's idea was we ought to be able to prove, then, that certain formulas cannot be derived within the system by those rules. Like, we can't derive the formula "p and not p." We can't derive a contradiction.

Think of this formal system as a formal game played with symbols. Then, what we want to show is this. The rules of this formal mathematical game do not allow us to derive the formula "p and not p." That would show—if we could prove that, if we could prove that formula could not be derived within the formal system representing mathematics—that would prove that there are no contradictions in mathematics and, therefore, that mathematics is consistent.

We can get a feeling for what Hilbert had in mind if we look at a very simple formal system with rules—very simple formal system with rules. Consider the game of tic-tac-toe. Suppose I tell you this is a supposed position in the game of tic-tac-toe. You know that this position cannot occur in tic-tac-toe. How do we know that? Because we know the rules of the game. Hilbert wanted to do the same thing for mathematics. He wanted to prove that in a formal system rich enough to include all of number theory (you could say all of arithmetic, if you prefer), no contradiction could occur. Once he had proved that mathematics couldn't contain contradictions, Hilbert believed mathematics would continue to progress, with the eventual goal of proving every result that was true in our axiomatized system—as

Descartes and Leibniz had dreamed. So, did Hilbert's idea work? Can we prove that mathematics is consistent, that contradictions can never occur? Unfortunately, no.

In 1931, Kurt Gödel proved 2 important things about any axiomatic system rich enough to include all of number theory. First, you'll never be able to prove every true result. Let me repeat that. Gödel proved that you'll never be able to prove every result that is true in your system.

Second, Gödel proved that one of the results that you can never prove is the result that says that the system is consistent. More precisely: You cannot prove the consistency of any mathematical system rich enough to include the known theory of numbers.

These are remarkable results. One more time—first, Gödel proved that there are propositions whose truth or falsity cannot be proved within the system. Any consistent mathematical system that is rich enough to include number theory is inherently incomplete.

Second, one of the propositions whose truth or falsity cannot be proved within the system is precisely the proposition that states that the system is consistent. What Gödel's proof means, then, is that we can't prove that arithmetic—let alone any more-complicated system—is consistent. So, Hilbert's dream is dead—Descartes's dream is dead.

Now listen to this implication: For 2000 years, mathematics has been the model—the subject—that convinces us that certainty is possible. But since Gödel's proof, as the existentialist philosopher William Barrett has put it: "Now there's no certainty anywhere—not even in mathematics." If all that isn't revolutionary enough, let me look at the fourth topic: the digital computer. People often think of computing as a technology and as part of engineering—but its backbone is mathematics, logic, and Descartes's "divide-and-conquer" approach to problem solving. A computer is a processor of symbolic information. Computers have transformed the economy—they have transformed communication, and they've also strongly influenced education in the modern world.

A personal testimony—in 1959, I had a job working at a national laboratory that had an IBM 709 computer, that was totally state of the art. The computer occupied a whole building. I needed to use it only rarely, which was just as well because you had to give it your data and your program on punched cards, and then you had to come back hours later for your printed results.

Programming—that's telling the computer what to do—was in a highly technical symbolic language. The language was very close to what the computer was actually doing. Still, it was great to go over there to use the computer, because its building had to be kept well air-conditioned, because it was before the transistor age and the computer ran on vacuum tubes.

Fast forward to today—I mean almost literally "today." As I wrote this lecture, I used a word processing program. No programming on my part needed—if I want to copy something, I don't need to know how the computer will do it; there's a simple command called "copy." I took a break from writing this lecture to get the latest political news over the Internet. Then, I e-mailed a colleague in Scotland about a manuscript by MacLaurin that we were both interested in. For all 3 of these activities, I used my laptop computer. It fits into a briefcase, and it doesn't seem to care what the room temperature is. It has a million times more memory and is a million times faster—literally. While this all was going on, my husband was paying our bills—all of the bills are generated by computers. Then, he went to pick up our car—something was wrong with it. The mechanic fixed it after checking out the engine on his computer. That was all in one afternoon. I'm sure you can think of examples like this from your own life.

Scientists now record their data on computers. They use computers to analyze the data. They use computers to set up very complicated mathematical models whose analysis is far beyond the speed of human thought. The mathematics of dynamical systems and fractals, for instance, would be unimaginable without the power of computers. The use of computers by artists, musicians, filmmakers, architects, and city planners has transformed the arts and the environment and has expanded the human imagination.

Computers and computing raise many social questions, notably about computers' effect on privacy, and on whether instantaneously available information—and misinformation—requires changing how we handle education. The existence of computers also has raised fundamental questions about what it means to be human. Can a computer think? Can computers develop consciousness? Descartes said that a machine could not think, but, of course, calculating machines were very primitive in the 17$^{th}$ century.

When I was teaching computing, I used to have a cartoon on my office door, with 2 scientists standing in front of a great big computer that had just printed something out. One scientist was reading the printout, and he says:

"I'll be darned. It says: 'I think, therefore I am.'" Can computers think? Will they ever be able to? The jury is still out on this question.

We ought to touch today on one more mind-bending idea: the idea of the infinite. In the 19<sup>th</sup> century, the German mathematician Georg Cantor worked out the first rigorous mathematical theory of the actual infinite. What do I mean, "actual infinite"? Aristotle had allowed things to be potentially infinite—like the list of natural numbers: We have 1, 2, 3, 4, 5, and so on, but no matter how large a number you give me, I can always give you a bigger one.

So, the list of numbers is infinite, but only potentially infinite, because I can never get to the end and have a completed infinite set with all the natural numbers. That's Aristotle's idea.

But partly because of the medieval idea of the infinity of God, partly because of the Renaissance idea of the infinite universe, and partly because dealing with infinite sets, like the set of all the points on a line segment, became necessary in mathematics after the invention of the calculus—because of all those things, mathematicians came to want and need the actual infinite. Cantor devised a theory of actually infinite sets.

Just to whet your appetite on this one—details can be found in the recommended books by George Gamow and Rosza Peter—Georg Cantor proved that the infinite set of natural numbers was of the same size as the infinite set of fractions. Really!

The infinite set of natural numbers 1, 2, 3, 4, and so on is the same size as the infinite set of all fractions: 3/5, 5/3, 355/113, and so on—seems like there should be more fractions, right? Cantor proved that those sets have the same number of elements in them. Cantor also proved that the infinite set of real numbers—that includes not only whole numbers and fractions, but irrational numbers as well, numbers like the square root of 2, and pi, the ratio of the circumference of a circle to its diameter—Cantor proved that the infinite set of real numbers is bigger than the infinite set of natural numbers. Bigger!

Does that seem very unlikely? It should. The same Kurt Gödel who proved that we can't prove the consistency of number theory, Gödel also proved that the introduction of the infinite, as mathematicians had been using it, does not by itself introduce inconsistencies into mathematics. So once more, proof succeeds in establishing a new truth—a truth that violates our expectations and that transcends our imagination.

I promised you a summing-up, so here it is. With the increasing role of mathematics everywhere in the modern world, it's more important, I think, than ever for everybody—not just mathematicians, and scientists, and philosophers—for everybody to understand how mathematics interacts with all areas of human life. In this course, I've tried to provide a historical perspective on that interaction. Even given the new mathematical examples that I introduced in this lecture, I stand by my initial contention: The branches of mathematics that have had by far the greatest impact on Western thought are probability and statistics, and geometry.

Mathematics is fun, and amazing, and creative—it's also powerful and influential. So, a crucially important question for the modern world is: How do we use mathematics and understand mathematics without pushing its power and authority beyond its proper domain? How can we use and understand mathematics in ways that enhance human existence? My hope is that this course has helped you get a handle on those questions, and also has inspired you to pursue these topics further. I thank you very much for your attention.

# Timeline

c. 1427.............................................Masaccio's *Trinity*.

1435 .............................................Leon Battista Alberti's Latin treatise on painting.

1455 .............................................Johannes Gutenberg prints the Bible.

c. 1480.............................................Piero della Francesca's *On Perspective in Painting*.

1492 .............................................Christopher Columbus's first voyage to the Americas.

c. 1495.............................................Leonardo da Vinci's *Last Supper*.

1517 .............................................Martin Luther's 95 Theses.

1543 .............................................Nicolaus Copernicus's *On the Revolutions of the Celestial Orbs*.

1591 .............................................François Viète introduces general symbolic notation in algebra.

c. 1600.............................................William Shakespeare's *Hamlet*.

1620 .............................................Francis Bacon's *Novum Organum*.

1637 .............................................René Descartes' *Discourse on Method*, *La Géometrie* (his book on analytic geometry, which was also independently invented by Pierre de Fermat).

1654 .............................................Beginning of Pascal-Fermat correspondence: birth of modern probability theory.

1675 .............................................Benedict de Spinoza's *Ethics Demonstrated in Geometrical Order*.

1684 .............................................Gottfried Wilhelm Leibniz's first publication of his differential calculus.

1687 .............................................Isaac Newton's *Principia* (*Mathematical Principles of Natural Philosophy*), includes Newton's laws of motion and his law of universal gravitation.

1850 .............................................. Second law of thermodynamics formulated by Clausius.

1854 .............................................. Riemann's "On the Hypotheses Which Lie at the Foundation of Geometry."

1855 .............................................. Death of Carl Friedrich Gauss.

1859 .............................................. Charles Darwin's *Origin of Species*.

1865 .............................................. Gregor Mendel's "Experiments on Plant Hybridization."

1870 .............................................. Hermann von Helmholtz's "On the Origin and Significance of Geometrical Axioms."

1871 .............................................. James Clerk Maxwell's *Theory of Heat*.

1872 .............................................. W. K. Clifford lectures on non-Euclidean geometry in England.

1898–1972 ..................................... Life of M. C. Escher.

1905 .............................................. Albert Einstein's special theory of relativity.

1913 .............................................. Niels Bohr's atomic theory.

1914–1918 ..................................... World War I.

1916 .............................................. Einstein's general theory of relativity.

1919 .............................................. Eclipse of the Sun: Measurements confirm Einstein's general theory of relativity.

1931 .............................................. Kurt Gödel's proof that the consistency of arithmetic cannot be proved.

1939–1945 ..................................... World War II.

1963 .............................................. Martin Luther King's "I have a dream" speech.

1972 .............................................. Edward Lorenz's MIT lecture on chaos and dynamical systems.

# Glossary

**alternate interior angles**: If 2 parallel lines are cut by a third line, the alternate interior angles are those within the space enclosed by the 2 parallels, with the top one on one side of the third line and the bottom one on the other side of the third line. For example, angle 1 and angle 2 in the following picture.

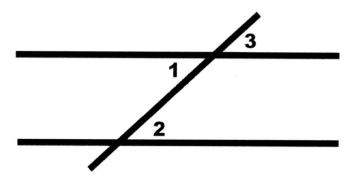

**argument**: In logic, an argument is a set of premises and a conclusion, where the premises are given as support for the conclusion.

**average expected value of some outcome**: The product of the probability of the outcome and the value of the outcome. Thus the average expected value of a coin-flipping game where you get $1 for heads and nothing for tails is 50¢, since the probability of a head is 1/2, and its value in the game is $1. If you played the game many, many times, you'd expect on average to win 50¢ each time you played.

**base rate of an event**: The frequency of events of that type. Before deciding that some event is unusual on the basis of how many times it occurs, like "employees taking 40% of their days off on Monday or Friday to stretch the weekend," one needs to look at the frequency: 40% of all workdays are Mondays or Fridays.

**bell curve**: If one graphs the frequency of some parameter on the vertical axis of a graph and the value of the parameter on the horizontal axis, and if the resulting distribution follows the exponential function $y = e^{-\frac{1}{2}(x^2)}$, resulting in a bell-shaped distribution, the graph is known as a bell curve, and the data is said to follow the "normal distribution." (The mean in the above equation is 0, but the equation can be adjusted so that the mean is any given number. It can also be "normalized" so that the area under the curve is 1 so that the curve represents a distribution of probabilities.) This distribution and the corresponding curve are important because many chance occurrences, and many human traits in populations, are distributed in accordance with it. The results of many psychological tests, including IQ scores, are set up to be distributed in this way as well. The bell curve is also called the normal curve or the Gaussian distribution.

**biased sample**: A sample of a population chosen in such a way that every individual in the population does not have the same probability of being chosen. The bias need not be deliberate.

**birthday problem**: The problem that asks, "How many people need to be in a room for there to be at least a 50% chance that 2 of them have the same birthday?"

**chaos**: *See* **dynamical systems and chaos**.

**choose function**: The mathematical function that gives the number of ways a set of $k$ objects can be chosen from a set of $n$ objects; also known as "$n$ choose $k$." For an example, see **combinations**.

**combinations**: The term "combinations" is used for the number of ways a set of $k$ objects can be chosen from a set of $n$ objects; it is also known as "$n$ choose $k$." Each set of $k$ objects is counted only once, since the set remains the same even if the order of the objects within the set is changed. For example, 7 choose 3 is $(7 \times 6 \times 5)/(3 \times 2 \times 1)$. That is, it's the number of permutations (arrangements) of 7 objects taken 3 at a time ($7 \times 6 \times 5$) corrected by being divided by the number of ways of arranging 3 objects ($3 \times 2 \times 1$), so that each set of 3 objects is counted only once. Also called the **choose function**.

**combinatorics**: Informally, the branch of mathematics that carries out "counting without counting": more precisely, the branch of mathematics that studies the enumeration, permutation, and combination of sets of objects and the mathematical properties these enumerations, permutations, and combinations have.

**converse of a proposition**: If a proposition has the form "If $p$, then $q$," where $p$ and $q$ are statements, then the converse (using the same $p$ and $q$) is the proposition of the form "If $q$, then $p$." For instance, if 2 straight lines are cut by a third line, the proposition "If the 2 initial straight lines are parallel, then the alternate interior angles formed are equal" is the converse of "If the alternate interior angles formed are equal, then the 2 initial straight lines are parallel." The truth of a statement does not necessarily imply the truth of its converse; if the converse is in fact true, this fact requires a separate demonstration.

**correlation**: Informally, 2 events that occur together are said to be correlated, with no assumption made about causality. More precisely, a correlation is a statistical relation between 2 or more variables such that systematic changes in the value of one of the variables are accompanied by systematic changes in the other variable(s).

**corresponding angles**: If 2 straight lines are cut by a third line, the angles on each of the original lines that stand in the same position with respect to the third line are corresponding angles. See angle 2 and angle 3 in the diagram following the definition of **alternate interior angles**.

**cost-benefit analysis**: A method of choosing between alternatives by examining whether the benefits outweigh the costs; both the probability of the alternatives and the value of the respective benefits and costs must be considered. Also called risk-benefit analysis, it is a part of decision theory.

**De Méré's problem**: If you throw 4 fair dice, what is the probability that at least one of them will show a 6 on its top face? This was one of the problems considered by Blaise Pascal in the early days of probability theory.

**deduction**: In logic, the process of reasoning in which the truth of a conclusion necessarily follows from the truth of the stated premises.

**demonstrative science**: A science that is logically structured, deriving the properties of its subject matter logically from explicitly stated definitions and assumptions and from previously proved propositions. Historically, Aristotle described how this should be done, and Euclid's *Elements of Geometry* provided an influential example.

**dynamical systems and chaos**: Dynamical systems are mathematical models of physical phenomena whose state (or instantaneous description) changes as a function of time. Some of these systems are said to be "chaotic," which is the term used to describe the behavior of systems that follow deterministic laws but appear random and unpredictable. Chaotic systems are said to be "sensitive to initial conditions," which means that small changes in those conditions can lead to quite different outcomes.

**entropy**: A measure of the disorder or randomness of a closed system. The entropy of a closed system tends to increase with time.

**fractal**: A complex geometric pattern exhibiting self-similarity, which means that small details of the pattern's structure, viewed at any scale, repeat elements of the overall pattern. The most famous fractal is the Mandelbrot set, which is the set of complex numbers $C$ for which a certain very simple function, $Z^2 + C$, calculated over and over again using its own output (starting with 0) as input, eventually converges on one or more constant values. Fractals arise in connection with nonlinear and chaotic systems and are used in computer modeling of regular and irregular patterns and structures in nature, including the growth of plants and the statistical patterns followed by seasonal weather.

**histogram**: A bar graph. More precisely, a way of graphically displaying numerical data in which the horizontal axis designates the value of the data, while the vertical height of a bar at each horizontal point shows how many pieces of data have that particular value.

**hypothetical argument**: An argument that has a major premise of the form "If $p$, then $q$." Plato argues that mathematics is not the highest form of knowledge, because it is hypothetical, since the postulates and axioms must be assumed in order for the subject to proceed.

**indirect proof**: *See* **proof by contradiction**.

**induction**: A form of reasoning that draws general conclusions based on individual instances, where the instances can be observations or the results of experiments. It yields probability, rather than certainty, and is thus to be contrasted with **deduction**.

**kinetic theory of gases**: The theory that explains the large-scale properties of gases—like pressure, temperature, or volume—by considering the gas as made up of molecules moving at different velocities.

**mean**: The sum of the individual values of each data point in a set of data, divided by the total number of data points. For instance, the mean of the set of numbers 1, 1, 2, 3, and 23 is $(1 + 1 + 2 + 3 + 23)/5 = 30/5 = 6$.

**measures of central tendency**: A general term for any measure purporting to locate the middle (center) of a set of data. By far the most common measures of central tendency are the **mean**, the **median**, and the **mode**.

**median**: Given a set of data put in numerical order, the median is the value that half of the data is above and half is below. For the set of numbers 1, 1, 2, 3, and 23, the median is 2. If the number of data points is even, the median is halfway between the 2 middle values. Thus for the set of numbers 1, 1, 2, 3, 10, 23, the median would be 2.5.

**mode**: Given a set of data, the mode is the most commonly encountered value. For the set of numbers 1, 1, 2, 3, and 23, the mode is 1.

**multiplication principle**: If one event can occur in *M* ways, and a second event can occur—independently of the first—in *N* ways, then the number of ways the 2 events can occur together is the product $M \times N$. The multiplication principle is the most powerful tool in elementary combinatorics.

*n* **choose** *k*: *See* **combinations** or **choose function**.

**normal curve**: *See* **bell curve**.

**ontological proof of God's existence**: A type of proof, medieval in origin, of the existence of God based on ideas about the nature of being (hence the term "ontological" from *ontos*, the Greek word for "being"). The method is often attributed to Anselm of Canterbury (1033–1109), but many other thinkers have attempted it.

**permutations**: The number of different ways of arranging a set of objects. For instance, the number of ways of arranging 7 objects, or the number of permutations of 7 objects, is $7 \times 6 \times 5 \times 4 \times 3 \times 2 \times 1 = 7!$ The number of ways of arranging 7 objects taken 3 at a time—that is, the number of permutations of 7 objects taken 3 at a time—is $7 \times 6 \times 5$.

**Playfair's axiom**: The geometric assumption that, given a straight line in a plane and a point in the same plane not lying on the original straight line, there is only one parallel to the original line through the given point. Though attributed to the Scottish mathematician John Playfair, this assumption goes back to the Greeks.

**principle of sufficient reason**: For anything that exists or happens, there must be a sufficient reason why it must be so and not otherwise. Usually attributed to G. W. Leibniz, but used in various forms by many other thinkers. It is related to concepts of symmetry, to the idea that space is the same in every direction, to the balancing of a lever with equal weights located at equal distances to a fulcrum, and to idealizations of probability such as the probability that a fair coin will come up heads exactly half the time.

**proof by contradiction**: A proof by contradiction establishes the truth of a given proposition by first assuming that the proposition is false, and from that drawing a conclusion that is contradictory to something that is known to be true. Also called indirect proof or, from Latin, reductio ad absurdum. An essential tool in mathematics and philosophy since the Greeks, including in Euclid's theory of parallels.

**quantum theory**: Part of modern physics explaining the nature and behavior of matter and energy on the atomic and subatomic levels, where energy and radiation occur in discrete amounts known as quanta. Because in this theory one cannot simultaneously measure position and momentum, or simultaneously measure both the wave-like and particle-like properties of light, quantum theory presents interesting challenges to traditional philosophies of science.

**random sample**: A sample of a population in which every element in the population has an equal probability of being chosen. Contrast with **biased sample**.

**reductio ad absurdum**: *See* **proof by contradiction**.

**reify**: To treat as a thing. For instance, to refer to the "average college student" as though she actually exists.

**right-skewed**: A skewed distribution is a distribution of data points that is not symmetrical; it is right-skewed if the tail that produces the asymmetry goes off to the right. Left-skewed is defined similarly, with the tail that produces the asymmetry going off to the left.

**sampling error**: More meaningfully called sampling uncertainty. The uncertainty resulting from measuring a random sample instead of measuring an entire population. The sampling error is, by convention, the margin of uncertainty such that the values for 19 out of 20 random samples taken from a population lie within plus or minus that margin of the value for the entire population. A good approximation to the sampling error in a reasonably large population is $\frac{1}{\sqrt{\text{sample size}}}$ .

**Septuagint**: The translation of the Hebrew Bible into Greek by Jewish scholars in Alexandria, Egypt, in the $3^{rd}$ century B.C.E.

**sound argument**: A logical argument that is not only valid but whose premises are true, and therefore whose conclusion must also be true.

**space**: We all think we know what space is, but it is not easy to define. Here are the ways in which some of the great minds studied in this course attempted to define it: Something extended in which physical bodies exist (Descartes, modern dictionaries). An a priori transcendental form of intuition in which we order our perceptions (Kant). Something absolute, which in its own nature, without relation to anything external, remains always similar and immovable; real or absolute motion takes place with respect to this absolute space (Newton). A fiction; all that really exist are bodies and the relations between them (Leibniz). A mathematical model (actually, of space-time) whose equations allow us to describe the physical universe; the equations for the model describe greater curvature near bodies with larger mass (Einstein).

**standard deviation**: A measure of the spread of a distribution. Formally, the square root of the **variance**. A distribution has a small standard deviation when the values of the data are grouped closely around the measures of central tendency, and a large one when they are spread widely. For instance, the distribution 1, 2, and 3 would have a considerably larger standard deviation than the distribution 1.9, 2.0, and 2.1. Standard deviation is represented by the symbol σ (sigma) and can be calculated by using the σ key on many electronic calculators.

**statistical thinking**: Quantitatively reasoning about populations, without focusing on any specific individuals.

**stratified sample**: A sample chosen by first dividing a population into groups (or strata, hence the name "stratified") based on common characteristics; in public-opinion polling, these groups often are chosen on the basis of characteristics thought to influence voting behavior, such as income level, gender, ethnicity, or education.

**syllogism**: A form of deductive reasoning consisting of a major premise, a minor premise, and a conclusion, where the premises share a term (the middle term). Syllogisms were classified by Aristotle in his major work on deductive logic, the *Prior Analytics*, part of his *Organon*.

**theorem**: The Greek word for a proved proposition in geometry.

**valid argument**: In logic, an argument is said to be valid if whenever the premises are true, the conclusion must be true. Arguments are valid by virtue of their form, not their content. If the form of an argument is a valid one, then if the conclusion is false, one of the premises must be false as well.

**variance**: A measure of the spread of a set of numerical data. More precisely, the variance is the average of the squares of the differences between the mean value for the population and each of the separate values.

**vortices**: Whirlpools in the ether, according to Descartes' physics. Such vortices were formed, according to Descartes, by large physical bodies like the Sun, and the vortices carry each planet in its orbit around the Sun.

**weighting a sample**: In weighting, the results from a sample are adjusted to make up for the sample's nonrepresentativeness. For instance, if a population is 50% male, 50% female, and the sample turns out to be 60% male, 40% female, multiplying the number of responses from females by 3/2 would weight the females' responses so that they would be equal in number to the males' responses.

# Biographical Notes

**Aristotle** (384–322 B.C.E.): Born in Stagira, Greece, Aristotle was a student of Plato. He and Plato were the 2 most influential philosophers of the ancient world. Aristotle's writings on ethics, politics, rhetoric, poetics, biology, physics, astronomy, natural history, logic, and metaphysics were all seminal works throughout the history of Western civilization. Important thinkers influenced by Aristotle ranged from Euclid to Darwin. His systematic work on deductive logic in the *Prior Analytics* and on the nature of demonstrative science in the *Posterior Analytics*, together known as the *Organon* (tool or instrument for investigation), laid the groundwork for subsequent work on logic and the philosophy of science.

**Bolyai, János** (1802–1860): Born in what was then Hungary, the son of the mathematician Farkas Bolyai, János studied at the imperial engineering academy in Vienna and then pursued a career in the army. Intellectually, he turned to mathematics, at first following his father's attempts to prove Euclid's parallel postulate. But in the 1820s, he realized that a different geometry could be constructed without the parallel postulate, and he triumphantly wrote his father, "I have created a new world out of nothing." When his father informed Gauss of János's achievement, Gauss responded "I cannot praise it … since to praise it would be to praise myself." Being scooped by Gauss disillusioned the young Bolyai, though János allowed his father to publish his work in 1832. It was rescued from obscurity only in the 1860s, but Bolyai ranks with Gauss and Lobachevsky as an independent codiscoverer of non-Euclidean geometry.

**Condorcet, Marie-Jean-Antoine-Nicolas Caritat, Marquis de** (1743–1794): Born in Ribemont, France, Condorcet moved to Paris, studying mathematics at the Collège de Navarre. He moved in circles that included leading mathematicians and philosophers, including d'Alembert, Bezout, Voltaire, and Lagrange. His mathematical work included contributions to differential equations, probability, and the applications of mathematics to society, especially in voting theory. He believed that scientific progress presaged a bright future for the human race. After the French Revolution of 1789, Condorcet became deeply involved in its politics, and when the Girondists were expelled from the Convention of 1792, he had to run for his life. While hiding from the authorities, he wrote the work for which he is best known, *Sketch for a Historical Picture of the Progress of the Human Mind* (1795). He was arrested and was found dead in his cell at Bourg-la-Reine prison.

**Descartes, René** (1596–1650): Born to an aristocratic family in France—his father was *conseiller* to the Parlement of Brittany—he was educated by the Jesuits at the College of La Flèche. He then struck out on his own, traveling, serving in the Bavarian army, and reading and thinking. His most famous philosophical work is the *Discourse on Method*, which included the argument "I think, therefore I am"; the method of analysis as a key to philosophy; and the notion of clear and distinct ideas as guarantors of truth. The *Discourse on Method* was published along with scientific discoveries he claimed to have made by his method: one on optics; one on meteorology; and the most famous, on geometry, which marked him as an independent coinventor of analytic geometry (as was Fermat). Even more influential in philosophy are his *Meditations*, and in physical science, his *Principles of Philosophy*, which included an anticipation of what we now call Newton's first law of motion and a materialistic theory of the Copernican solar system. In 1649, Descartes became tutor to Queen Christina of Sweden. He died in Stockholm in 1650.

**Euclid** (4th century B.C.E.): Little is known about Euclid's life, save that he worked and taught in Alexandria, Egypt—perhaps he founded the great mathematical school there—and that he probably worked earlier than Archimedes and certainly worked later than Aristotle. Platonists claim that Euclid studied in Athens with Plato's successors, but there seems to be no evidence to support this. His work, though, was—and still is—more influential than that of any other ancient mathematician. Best known are the 13 books that make up Euclid's *Elements of Geometry*. The *Elements* is the ancestor of all high school geometry texts. Euclid's theory of parallels; his proof of the Pythagorean theorem (Book 1, Proposition 47); and his proofs that there are exactly 5 regular solids, that there are an infinite number of primes, and that the area of a circle is proportional to the square of its diameter all are still taught today.

**Fermat, Pierre de** (1601–1655): Fermat was born in France and entered the legal profession, getting a degree in Civil Law from the University of Orleans in 1631. He worked as a lawyer and *parlementaire* in Toulouse. So he was not formally a professional mathematician, but in terms of his contributions, he was greater than almost all the professionals. He was an independent coinventor of analytic geometry (as was Descartes). Fermat anticipated many of the discoveries of the calculus and formulated Fermat's principle in optics. He made important contributions to number theory, including a method of proof called infinite descent, which is an inverse version of what is now called mathematical induction. A conjecture he made, known as Fermat's last theorem, was considered of great importance by mathematicians for centuries, and work on it led to a great deal of mathematical progress (it was finally proved by Andrew Wiles in 1995). In his correspondence with Pascal, Fermat helped lay the foundations of the modern theory of probability.

**Gauss, Carl Friedrich** (1777–1855): Born in Brunswick, Germany, Gauss was a child prodigy, and the Duke of Brunswick recognized this and offered him support. Considered one of the greatest mathematicians and physicists in history, Gauss worked in arithmetic, number theory, algebra, analysis, geometry, probability and statistics, astronomy, geodesy, geomagnetism, mechanics, optics, and physics. His manuscripts make clear that he was the first to conceive of a non-Euclidean geometry, though he didn't publish his ideas or work them out in full detail. He was director of the observatory at Göttingen from the time he was 30 until his death. The French Revolution, the Napoleonic wars in Germany, and the democratic revolutions of 1848 did not produce a quiet life, but he continued his work on both the pure and applied sciences nevertheless. It was Gauss who chose the topic—the foundations of geometry—for Riemann's inaugural lecture in 1854 at Göttingen, and Gauss had the pleasure of hearing that lecture shortly before his death in 1855.

**Gould, Stephen Jay** (1941–2002): An American evolutionary theorist, paleontologist, and historian of science, Gould is known to the general public for his popular writings on science. He was born in New York City and was first inspired to become a scientist by the dinosaur exhibits in that city's Natural History Museum. In biology proper, his best-known work has been on the theory of punctuated equilibrium. Gould was both a professor at Harvard and a member of the U.S. National Academy of Sciences. Among his works for the general public, perhaps the best known is his 1981 book *The Mismeasure of Man*, a statistically sophisticated critique of arguments for the genetic inheritance of intelligence. The second edition (1996) included a critical review of Herrnstein and Murray's book *The Bell Curve*.

**Halley, Edmond** (1656–1742): Halley studied at Oxford but also was a seaman. His voyages included mapping the stars of the Southern Hemisphere and determining the variation of the Earth's magnetic field. Halley both persuaded Newton to write the *Principia* and underwrote its publication expenses. The mortality tables for Breslau, Germany, that Halley compiled helped found actuarial science in the 18th century. He is also known for his study of comets, including Halley's comet, and plotting their orbits. He became Astronomer Royal in 1720 and later died at Greenwich.

**Helmholtz, Hermann von** (1821–1894): Born in Potsdam, Germany, Helmholtz went to Berlin to study medicine, but he also went to lectures on chemistry and physiology and studied mathematics and philosophy on his own. He is best known as an independent codiscoverer of the law of conservation of energy. But he also made important contributions to the physiology of perception, especially in the theory of how we see color and in the perception of sound. His work on the philosophy of science included discussion of ideas about space, and translations of his work "On the Origin and Significance of Geometrical Axioms" introduced non-Euclidean geometry to the English-speaking world. A professor first at Königsberg, then Bonn, then Heidelberg, and then Berlin, Helmholtz is recognized as one of the foremost scientists of the 19th century.

**Hippocrates of Chios** (c. 470–c. 410 B.C.E.): Born in Chios, he needs to be distinguished from his contemporary, the well-known physician Hippocrates of Cos. Hippocrates of Chios worked in Athens, which until the founding of Alexandria was the leading center of Greek mathematical research. He is best known for being the first writer of an "Elements of Geometry," which, like all the "Elements" that preceded Euclid's, has been lost. Hippocrates, recognizing the difference between theorems that are interesting just for themselves and those whose truth leads to the proof of later theorems, put together a work that scholars believe contained much of what is now in Books 1 and 2 of Euclid's *Elements* as well as important theorems about circles. Hippocrates of Chios also worked on 2 of the famous problems of Greek mathematics—duplicating the cube and squaring the circle—and contributed to astronomy.

**Jefferson, Thomas** (1743–1826): Jefferson, who is best known as the principal author of the American Declaration of Independence, was born in Shadwell, Virginia. His father was a surveyor, his mother a member of the prestigious Randolph family. Jefferson, who attended the College of William and Mary, knew more of the mathematics of his time than did any other U.S. president. He designed a plow, using calculus to figure out the shape that would slide most smoothly through the soil, and contributed to voting theory, especially the problem of fairly apportioning representatives in Congress by population. Through his *Notes on the State of Virginia*, he became recognized as a scientific observer of international repute. He was ambassador to France in the 1780s, secretary of state under George Washington, vice president of the United States from 1797 to 1800, and the third president of the United States beginning in 1801. Jefferson died at Monticello, Virginia, on July 4, 1826. His tombstone, at his request, identifies him as the author of the Declaration of Independence, the author of the Statute of Virginia for Religious Freedom, and the founder of the University of Virginia.

**Kant, Immanuel** (1724–1804): Born in Königsberg, Prussia (now Kaliningrad, Russia), and living there all his life, Kant was the premier philosopher of the 18th century. He wrote a short pamphlet in 1784, "What is Enlightenment?" that beautifully summed up the leading ideas of the age, writing, "*Sapere Aude*! [Dare to know]. Have the courage to use your own understanding!—that is the motto of enlightenment." Kant's best-known work is the *Critique of Pure Reason* (1781; revised edition, 1787), which argued that we do not know things as they really are, but we reason from phenomena mediated through intellectual categories like causality. Our sense perceptions, according to Kant, are ordered in our unique intuitions of space and time. Other important Kantian works are the *Prolegomena to Any Future Metaphysics* (1783); the *Critique of Practical Reason* (1788), which is about ethics; and the *Critique of Judgment* (1790), concerning aesthetics. Kant's influence on subsequent philosophy has been immense.

**Lagrange, Joseph-Louis** (1736–1813): Born in Turin, Italy, Lagrange, at the age of 18, discovered some of the basic results of the calculus of variations. He taught at the Artillery School in Turin and helped found the Academy of Sciences there. In 1766, he was named director of the mathematics section of the Berlin Academy of Sciences. Lagrange worked on important problems in astronomy and celestial mechanics, including the 3-body problem. He also did fundamental work on the theory of equations and devised, for permutation groups, what is still called Lagrange's theorem. He moved to Paris in 1786, where he published his classic *Analytical Mechanics* (1788). After the Revolution, he became a professor at the newly founded Ecole Polytechnique, where his lectures on the foundations of the calculus helped lead to its rigorization. Throughout his career, he championed pure analysis over geometric intuition and advocated the greatest possible generality of concepts and methods. When he died, his body was interred in the Panthéon in Paris, and his papers, at the order of Napoleon, were deposited in the library of the Institut de France.

**Laplace, Pierre-Simon** (1749–1827): Born in Normandy, France, Laplace began as a student of theology, but when he went to Paris, he impressed d'Alembert, who got the young Laplace a position as professor of mathematics at the Ecole Militaire. He was elected to the Académie des Sciences and later became a professor at the Ecole Polytechnique. Although his work in mathematical analysis was important and influential, Laplace is best known for his celestial mechanics. He was an independent coinventor (as was Kant) of the nebular hypothesis for the origin of the solar system and proved the stability of the solar system with very sophisticated mathematics. He was also a prominent contributor to the theory of probability (especially the theory of errors) and its application to astronomy as well as a contributor to the philosophy of probability.

**Leibniz, Gottfried Wilhelm** (1646–1716): Born in Leipzig, Germany, Leibniz contributed to a wide range of subjects, from symbolic logic to the technology of mining. He worked as a diplomat, traveling all over Europe, and wrote on history and the law. Within mathematics, Leibniz championed the use of powerful heuristic symbolic notation, hoping that such notation could be used in all areas of thought to find truth. More mundanely, he devised the notation still used in calculus today for the differential and the integral. Within philosophy, Leibniz is known for his formulation of the principle of sufficient reason. He was involved in a dispute with the Newtonians over the role of God in the universe, and he opposed the Newtonian doctrine of absolute space and time. Mathematicians know Leibniz best as an independent coinventor of the differential and integral calculus (as was Newton). He continued his political, legal, and historical work throughout his life.

**Lobachevsky, Nikolai Ivanovich** (1792–1856): Born in Nizhni Novgorod (now Gorki), Russia, Lobachevsky studied mathematics at Kazan University under Martin Bartels, a friend of Carl Friedrich Gauss. Lobachevsky became a professor there and later became dean, librarian, and then rector of that university. Lobachevsky was one of the 3 original independent discoverers of non-Euclidean geometry (along with János Bolyai and Gauss). At first he tried to prove Euclid's fifth postulate, but he realized in 1821 that a geometry that denied that postulate was not inherently contradictory, and so he decided to create such a geometry. Lobachevsky first reported on his non-Euclidean geometry to the Kazan University faculty in 1826, calling it "imaginary geometry," and in 1829, he became the first to publish a non-Euclidean geometry. His subsequent publication in German in 1840 introduced the subject to a wider European audience. The geometry that denies Euclid's fifth postulate and allows more than one parallel to a given line through an outside point is called Lobachevskian geometry.

**Maxwell, James Clerk** (1831–1879): One of the greatest physicists in history, Maxwell was born in Edinburgh, Scotland. He studied first at Edinburgh, then at Cambridge, receiving his degree from Trinity College in

1854, where he then became a fellow. He gave an elegant theory of the stability of the rings of Saturn under the assumption that they were made of small, solid particles—a result widely accepted (and recently experimentally confirmed by the *Voyager* spacecraft). He also was an independent coinventor (as was Boltzmann) of the statistically based kinetic theory of gases and gave an influential statistical interpretation of the second law of thermodynamics. Maxwell became the Cavendish Professor at Cambridge and helped set up the famous Cavendish laboratory there. He is best known for his formulation of Maxwell's equations, which encapsulate the relationship between electricity, magnetism, and electromagnetic vibrations such as light.

**Newton, Isaac** (1642–1727): Born in Woolsthorpe, England, Newton studied at Cambridge University. He was an independent coinventor of the calculus (as was Leibniz) and the founder of modern experimental optics. His *Mathematical Principles of Natural Philosophy* (1687), usually known by the abbreviated Latin title *Principia*, gave a mathematical theory of the universe that explained, and quantitatively predicted, phenomena ranging from the fall of an apple, the shape of the Earth, and the tides of the oceans to the orbits of the planets and their satellites. The *Principia* also established the ideas of absolute (Euclidean) space and absolute time, essential to modern mathematical physics until the time of Einstein. Knighted by William III, Newton was buried in Westminster Abbey with the honors usually accorded to royalty.

**Pascal, Blaise** (1623–1662): Born in Clermont-Ferrand, France, Pascal began his mathematical studies with Euclid's *Elements*. He became interested in projective geometry when he was 17. He wrote on the conic sections, devised a calculating machine that could do multiplication (for this reason, in the 20th century a computer language, Pascal, was named after him), and studied hydraulics and the laws of fluids and gases. By means of the idea of "indivisibles," he worked on areas, volumes, centers of gravity, and the lengths of curved lines, thus becoming a forerunner of the inventors of the calculus. He and Fermat exchanged letters about problems of gambling and thereby invented modern probability theory. Pascal also worked on problems in combinatorics, including the Pascal triangle of binomial coefficients, and developed a sophisticated version of the proof method now called mathematical induction. He wrote on theology as well. His *Pensées* (short, sometimes fragmentary thoughts on philosophical and religious matters) have become famous, and the one called "Infinity–Nothing: The Wager" includes the first real cost-benefit argument in history.

**Piero della Francesca** (c. 1415–1492): Born in Borgo Sansepolcro, Italy, the son of a dealer in leather and dyes, Piero studied mathematics but preferred painting and was apprenticed to a local painter. Piero wrote several treatises on mathematics, of which 3 survive: an elementary work on arithmetic, geometry, and polyhedra; *The Short Book on the Five Regular Solids*; and the most famous, *De prospectiva pingendi* ("On perspective for painting"). This work is the first treatise devoted to the geometry of perspective. Although none of Piero's mathematical work was published under his own name in the Renaissance, it circulated widely in manuscript and became influential through its incorporation into the works of others. Piero also was an accomplished artist, and his works are part of the great legacy of Renaissance Italian painting.

**Plato** (c. 427–c. 347 B.C.E.): Plato is widely recognized as the founding father of the entire Western philosophical tradition. A native of Athens and a pupil of Socrates, Plato wrote a series of dialogues addressing important philosophical questions. Most famous is his *Republic*, envisioning an ideal state ruled by philosopher-kings. In this work, Plato discusses the nature of justice; the best type of education—with mathematics at its heart; and his theory of eternal and unchanging forms, of which objects in the visible and tangible world are only imperfect images. Among Plato's other dialogues is the *Timaeus*, in which the 5 regular solids, often called Platonic solids, form a mathematical basis for the creation of the world. He founded a school in Athens, called the Academy, which survived until closed by the Roman emperor Justinian in the $6^{th}$ century C.E. It is said that the motto "Let no one ignorant of geometry enter here" was written over the doorway.

**Quetelet, Lambert Adolphe Jacques** (1796–1874): Born in Ghent, Belgium, Quetelet in 1819 received the first doctorate given by the new University of Ghent. He went to Paris in 1823 to study astronomy and then became astronomer at the Royal Observatory in Brussels. He became interested in statistics, first from the use of probability in astronomy and then in the study of human beings in society. His work *Sur l'homme* (*On Man*; 1835) introduced the idea of the "average man." In later works, he showed how the frequency of various human traits, like height, was distributed in the human population in the bell-shaped curve characteristic of the theory of errors in measurement. He also gave many examples of the stability of social statistics, including examples like suicides, which, previously thought of as individual free choices, seemed to occur at virtually the same rate every year. Quetelet explained this constancy of rates by postulating that it was produced by constant social causes. His work played a crucial role in developing statistical thinking, especially about society.

**Riemann, Bernhard** (1826–1866): Born in Breselenz, a small town near Dannenberg, Germany, Riemann attended the University of Göttingen, at first to study theology and philology, but eventually concentrating on mathematics. In 1847, he moved to Berlin, studying with Jacobi and Dirichlet. Returning to Göttingen, he wrote his thesis on complex functions and Riemann surfaces in 1851. For his inaugural lecture, Riemann proposed 3 topics, and Gauss made an important choice, asking that Riemann speak on the foundations of geometry. Riemann's lecture was a significant contribution to the general study of geometries on generalized spaces called manifolds. Riemann also made major contributions to complex function theory, to algebraic functions, to differential geometry, and to the theory of integration in calculus. Proving the Riemann hypothesis is still a major unsolved problem in mathematics. And Riemannian geometry was essential in Einstein's general theory of relativity.

**Spinoza, Benedict de** (a.k.a. **Baruch Spinoza**; 1632–1677): Spinoza was born into Amsterdam's Portuguese Jewish community. He was a gifted student but soon questioned the prevailing religious ideas of both Judaism and Christianity, using Cartesian rationalism to reject the idea that the Bible was the divine word. His *Ethics* argues that human happiness and well-being do not come from a life pursuing material wealth or passion, or from accepting what he called the "superstitions that pass as religion," but from the life of reason. The *Ethics*, whose full title is *Ethics Demonstrated in Geometrical Order*, was presented in the definition-axiom-proposition form characteristic of mathematical works. Spinoza's rational critique of religion and defense of freedom of thought in his *Theological-Political Treatise* has also been influential.

**Voltaire** (a.k.a. **François Marie Arouet**; 1694–1778): Born in Paris, Voltaire was the quintessential intellectual of the French Enlightenment, giving readable and eloquent accounts of its leading ideas. He began his education at the Jesuit College of Louis-le-Grand, ironic in the light of his later attacks on conventional religion. Voltaire had a varied and colorful career. He wrote somewhat scandalous plays and poetry and was briefly imprisoned in the Bastille. He later visited England and was impressed at the respect that the English accorded to Newton. Voltaire fell in love with a married aristocratic woman, Madame du Châtelet, lived with her, and collaborated with her to write a commentary on Newton's physics. Voltaire also wrote more plays, became royal historiographer for a while through the influence of Madame de Pompadour, and moved to Berlin, where he began to write his *Philosophical Dictionary*. Objecting to the Leibnizian doctrine that this was the "best of all possible worlds," Voltaire wrote the poem "The Lisbon Earthquake" and the famous story *Candide*. He also wrote serious works of history. However, his freethinking got him into trouble, and his *Natural Law* was burned by the public hangman. He is wonderfully quotable. For instance: "This is the character of truth: it is of all time, it is for all men, it has only to show itself to be recognized, and one cannot argue against it. A long dispute means that both parties are wrong."

# Bibliography

Andersen, Kirsti. *The Geometry of an Art: The History of the Mathematical Theory of Perspective from Alberti to Monge.* New York: Springer, 2007. An immensely learned and detailed account (750 pages), with lots of illustrations, of the geometry of perspective from the Renaissance to the early 19<sup>th</sup> century. For those deeply interested in the topic.

Arendt, Hannah. *Eichmann in Jerusalem.* Rev. ed. New York: Viking Press, 1965. Reprint, New York: Penguin Classics, 2006. The classic account of the trial of Adolf Eichmann, architect of the Holocaust. Originally published in the *New Yorker* in 1963, Arendt's book is famous for its subtitle, *A Report on the Banality of Evil.* Without needing to subscribe to all of Arendt's thesis, a participant in this course will find interesting Arendt's rejecting the statistical view of "mass man" in favor of the responsibility of individuals.

Aristotle. *Prior Analytics.* Cambridge, MA: Harvard University Press, 2002. This is the deductive logic from Aristotle's *Organon.* He treats all the forms of the syllogism. This has been unbelievably influential, though it's tough going to read on one's own.

Armitage, David. *The Declaration of Independence: A Global History.* Cambridge, MA: Harvard University Press, 2007. Treats the influence of the Declaration of Independence of the United States in a world context, addressing both the non-American roots of the Declaration and the influence of the Declaration on the national movements of other, later states.

Barzun, Jacques. *Teacher in America.* London: Little, Brown, 1945. An unabashedly old-fashioned book about education by an iconoclastic but appealing cultural historian from Columbia.

Bayley, Mel. "*Hard Times* and Statistics." *British Society for the History of Mathematics Bulletin* (2007): 92–103. A fine article on the statistical aspects of Dickens's *Hard Times.*

Becker, Carl L. *The Declaration of Independence: A Study in the History of Political Ideas.* New York: Harcourt, Brace, 1922. Becker treats the ideas in the Declaration of Independence in their historical context, emphasizing the Declaration's origins in classical political philosophy, especially that of John Locke.

Bloom, Alfred. *The Linguistic Shaping of Thought: A Study of the Impact of Language on Thinking in China and the West.* Mahwah, NJ: Lawrence Erlbaum, 1981. An influential account of the way the differences in language in these societies may affect what it is possible to think about. A provocative study.

Blum, Harold F. *Time's Arrow and Evolution*. New York: Harper Torchbook, 1962. A popular introduction to the relationship between evolutionary biology and time, including discussion of the second law of thermodynamics.

Bonola, Roberto. *Non-Euclidean Geometry*. New York: Dover, 1955. A great geometer looks at original sources in the history of the discovery and development of non-Euclidean geometry. Mathematically sophisticated.

Brown, James Robert. *Philosophy of Mathematics: A Contemporary Introduction to the World of Proofs and Pictures*. 2nd ed. New York: Routledge, 2008. A good, though not wholly elementary, introduction to modern views on the philosophy of mathematics.

Chomsky, Noam. "The Fallacy of Richard Herrnstein's IQ." *Cognition* 1 (1973): 285–298. Reprinted in *Darwin*, 2nd ed. Edited by P. Appleman. New York: W. W. Norton, 1979. A provocative attack on the reification of average IQ, notable for its comparison of the different ways race and height have been treated in their statistically similar relation to IQ.

Cohen, I. Bernard. *Science and the Founding Fathers*. New York: W. W. Norton, 1995. A careful and thoughtful study of the actual scientific background of such men as John Adams, James Madison, and Thomas Jefferson.

Curley, Edwin. *Behind the Geometrical Method: A Reading of Spinoza's "Ethics."* Princeton, NJ: Princeton University Press, 1988. For those who want a serious and challenging look at the ethical ideas in Spinoza's work.

Davis, Philip J., and Reuben Hersh. *The Mathematical Experience*. Rev. ed. New York: Birkhäuser, 1995. A wonderful and light-hearted approach to the social and philosophical issues surrounding modern mathematics and mathematical practice. Written for nonmathematicians.

De Condorcet, Antoine-Nicola. *Sketch for a Historical Picture of the Progress of the Human Mind*. 1796. Reprint, New York: Noonday Press, 1955. This is the outstanding 18th-century French work advocating the perpetual progress of the human race on the basis of advances in mathematics, science, and technology.

Descartes, René. *Discourse on Method*. 1637 (any edition or translation). Descartes' most famous work, including "I think, therefore I am" and the rules for his method for finding truth in the sciences.

———. *Meditations on First Philosophy*. 1641 (any edition or translation). Thought by many to be Descartes' most significant and profound philosophical work, the *Meditations* concern God and the soul, and also matter and space.

Dickens, Charles. *Hard Times.* 1854 (any edition). A novel about the Industrial Revolution and the statistical view of life, with Dickens (of course) coming down instead on the side of humane conduct.

Dostoyevsky, Fyodor. *Notes from Underground.* 1864 (any edition or translation). Dostoyevsky's underground man has inspired much later writing, from the existentialists to Ralph Ellison's *Invisible Man*, by championing human free choice over all rational calculation.

Einstein, Albert, and Leopold Infeld. *The Evolution of Physics.* New York: Simon and Schuster, 1938. A short history of physics by 2 famous physicists, especially authoritative in its popular account of relativity.

Feingold, Mordechai. *The Newtonian Moment: Isaac Newton and the Making of Modern Culture.* New York: New York Public Library, 2004. Lavishly illustrated account of the life and work of Newton and its relationship to the culture of the Enlightenment; the book is based on a major exhibition at the New York Public Library.

Field, J. V. *The Invention of Infinity: Mathematics and Art in the Renaissance.* Oxford: Oxford University Press, 1997. An outstanding work of scholarly exposition, especially noteworthy for its account of the geometric background of Renaissance perspective and its treatment, including manuscript sources, of Piero della Francesca.

Fowler, David. *The Mathematics of Plato's Academy: A New Reconstruction.* 2$^{nd}$ ed. Oxford: Clarendon Press, 1990. A scholarly treatment of the subject, with a sophisticated view of the mathematics as well as the philosophy and history. One needs to know a fair amount of mathematics to deal with the main argument, which at some points disagrees with that of Wilbur Knorr's book.

Friedman, Michael. *Kant and the Exact Sciences.* Cambridge, MA: Harvard University Press, 1992. A superb and sophisticated philosophical and historical analysis of its topic.

Gamow, George. *One, Two, Three—Infinity: Facts and Speculations of Science.* New York: Dover, 1988. A nice popular account of a number of topics in mathematics and science, including the proofs that there are more real numbers than rational ones, but as many whole numbers as there are rational ones; also briefly introduces non-Euclidean geometry and its relationship to general relativity. Readable and interesting.

Gaukroger, Stephen. *Descartes: An Intellectual Biography*. Oxford: Oxford University Press, 1995. Comprehensive and based on good scholarship.

Gawiser, Sheldon R., and G. Evans Witt. *A Journalist's Guide to Public Opinion Polls*. Westport, CT: Greenwood, 1994. Easy to read, assuming no mathematical or statistical background, this book is an extremely valuable guide to the topic for its intended audience.

Gay, Peter, ed. *The Enlightenment: A Comprehensive Anthology*. New York: Simon and Schuster, 1973. This course's bibliography needs an anthology of writings from the 18th century exhibiting the role science played in the century's attitudes toward progress, universal agreement, religion, and various philosophical questions, and Gay's book fills the bill. It is a very good anthology with valuable scholarly commentary.

Gillispie, Charles Coulston, ed. "Quetelet." In *Dictionary of Scientific Biography*, vol. 11. New York: Scribner, 1970–1980.

Gleick, James. *Chaos: Making a New Science*. New York: Viking Press, 1987. A very good popularization of the ideas of chaos theory, dynamical systems, and fractals.

Gould, Stephen Jay. "The Median Isn't the Message." In *The Richness of Life: The Essential Stephen Jay Gould*, edited by Steven Rose. New York: Norton, 2007. Also in Stephen Jay Gould, *Bully for Brontosaurus: Reflections in Natural History*. New York: Norton, 1992. Excellent essay making the point that things like the median are statistical abstractions and that one must always remember that the variation, the individuals involved, are what truly matters.

———. *The Mismeasure of Man*. New York: W. W. Norton, 1996. The book as a whole is a historical study of various ways of measuring, and mismeasuring, human intelligence, informed by Gould's own research on statistics in biology, and by his view that the reification of statistical abstractions is misleading in the extreme. He argues strongly against the views of Herrnstein and Murray in *The Bell Curve*.

Grabiner, Judith V. "The Centrality of Mathematics in the History of Western Thought." *Mathematics Magazine* 61 (1988): 220–230. This article follows, albeit briefly, much of the core of this Teaching Company course.

———. "Descartes and Problem Solving." *Mathematics Magazine* 68 (1995): 83–97. An article arguing that Descartes' analytic geometry is not a new subject but a new method of solving problems, some from the ancient world and some new.

————. "The Use and Abuse of Statistics in the 'Real World.'" *Skeptic* (Summer 1992): 14–21. A short account of the topic, using some of the examples found in this Teaching Company course.

————. "Why Did Lagrange 'Prove' the Parallel Postulate?" *American Mathematical Monthly* 116 (January 2009): 3–18. An account of the manuscript in which Lagrange gives his supposed proof of the parallel postulate, in the form of Playfair's axiom, the article also sets the proof in historical context.

Gray, Jeremy. *Ideas of Space: Euclidean, Non-Euclidean and Relativistic.* 2nd ed. Oxford: Clarendon Press, 1989. The best book on the history of non-Euclidean geometry. Some of the material is technical and requires knowledge of geometry beyond that of a high school course, but the book is nonetheless important and rewarding.

*Grimshaw v. Ford Motor Co.*, 1 19 Cal. App. 3d 757, 174 Cal. Rptr. 348 (1981). The court decision in the Ford Pinto case, often used to illustrate the way cost-benefit analysis can be misused.

Hacking, Ian. *An Introduction to Probability and Inductive Logic.* Cambridge: Cambridge University Press, 2001. Though a textbook, this is written in such a user-friendly fashion that it does not seem like one. It begins with a set of examples that are fun to think about and explains a lot of very sophisticated concepts using ordinary language. Philosophically sophisticated but still quite readable. It contains a set of problems, and solutions (not just answers) are given in the back of this 300-page book.

Heath, Thomas L. *Mathematics in Aristotle.* Oxford: Clarendon Press, 1949. An older, but still valuable, scholarly work, surveying Aristotle's use of mathematics and what he says about it.

————, ed. *The Thirteen Books of Euclid's Elements.* 3 vols. New York: Dover, 1956. This cheap edition of Euclid is the way to acquaint oneself with the real thing, albeit in English and not informed by more recent scholarship such as that of Wilbur Knorr. Volume 1 contains both Book 1 of the *Elements*, the book covered in the course, and Book 2, about areas. Volumes 2 and 3 contain the rest of the *Elements*, with Volume 2 containing Books 3–9 and Volume 3 containing Books 10–13. Heath's edition of Euclid is also the source for Professor David Joyce's website that contains all of the *Elements* and some commentary: http://aleph0.clarku.edu/~djoyce/java/elements/toc.html.

Henderson, Linda Dalrymple. *The Fourth Dimension and Non-Euclidean Geometry in Modern Art.* Princeton, NJ: Princeton University Press, 1983.

Its 375 pages of text, 100 pages of bibliography, and more than 100 images make this insightful though challenging book the major source on its topic.

Herrnstein, Richard J., and Charles Murray. *The Bell Curve*. New York: Free Press, 1994. A provocative and controversial work that argues that a substantial part of IQ is inherited and that racial differences in IQ are not due just to environment. Contains many detailed statistical arguments. See also the works challenging its thesis in the Jacoby-Lieberman anthology and the introduction to the second edition of Gould's *Mismeasure of Man*.

Huff, Darrell. *How to Lie with Statistics*. New York: W. W. Norton, 1993. Absolutely the best popular introduction to evaluating statistical arguments ever written. Elementary, short, immensely readable, and illustrated with amusing cartoons.

Jacoby, Russell, and Naomi Lieberman, eds. *The Bell Curve Debate*. New York: Times Books, 1995. An anthology of articles responding to Herrnstein and Murray's *The Bell Curve*. Reasonably challenging to read, but valuable.

Jammer, Max. "Entropy." In *Dictionary of the History of Ideas*, vol. 2, 112–120. New York: Scribner's, 1983. By a leading philosopher who has also written on space and relativity, this is out of print but is available online at http://etext.lib.virginia.edu/cgi-local/DHI/dhi.cgi?id=dv2-12.

Kant, Immanuel. *Critique of Pure Reason*. 1781, rev. 1787 (any edition or translation). This is Kant's major work on pure reason; for those deeply interested, the actual text is worth grappling with.

———. *Prolegomena to Any Future Metaphysics*. 1783 (any edition or translation). Kant's explanation of what questions need to be answered before metaphysics is possible, notably "Are there any judgments independent of experience whose truth nevertheless does not appear merely by analyzing the terms in the judgment?" and "If there are, how can we tell if these judgments are true or false?"

Katz, Victor J. *A History of Mathematics: An Introduction*. 4th ed. Reading, MA: Addison-Wesley, 2008. Without a doubt, the very best history of mathematics available, covering the Western tradition in depth and many other cultures as well. Up-to-date in scholarship; clear and detailed in its mathematically sophisticated explanations. Some knowledge of mathematics is required for most of this book, and calculus and more are needed for the history of the modern period. Excellent bibliography; includes pronunciation of mathematicians' names in the index.

————. *A History of Mathematics: Brief Edition*. Boston: Pearson, 2004. A shorter (525 pages of text) version of Katz's definitive *History*. A bit more readable than the longer work, with good coverage of most topics in the longer work, and still has mathematical prerequisites.

Kemp, Martin. *The Science of Art: Optical Themes in Western Art from Brunelleschi to Seurat*. New Haven, CT: Yale University Press, 1990. From the Renaissance to the early 20th century, artists interacted with geometers, designers of instruments for representing perspective, and those knowledgeable about theories of color and vision. Thorough, documented, with over 560 illustrations, this is an excellent treatise on the topics covered.

Kitto, H. D. F. *The Greeks*. London: Penguin, 1950. A nice, short, readable introduction to the culture of the ancient Greeks, though obviously does not include more recent scholarship.

Klein, Jacob. *Commentary on Plato's "Meno."* Chicago: University of Chicago Press, 1989. A historically learned and mathematically informed commentary on the *Meno*.

Kline, Morris. *Mathematics in Western Culture*. New York: Galaxy Books, Oxford University Press, 1964. A magisterial work, addressing many of the topics in this Teaching Company course in its 472 pages. Though it does not reflect scholarship since the 1960s and reserves a good deal of its sympathies for those who share the modern Western worldview, it orients the reader well and contains many insights.

Knorr, Wilbur J. *The Evolution of the Euclidean Elements*. Dordrecht, Netherlands: Reidel, 1974. A challenging scholarly work that treats the historical origins of Euclid's work in full mathematical and linguistic detail. As with Fowler's book, one needs to know a fair amount of mathematics to master the argument.

Koerner, S. *Kant*. London: Penguin, 1955. A good short introduction to Kant's philosophy. It requires careful reading, but the subject is clearly presented.

Koerner, Tom. *The Pleasures of Counting*. Cambridge: Cambridge University Press, 1996. A nice introduction to many aspects of modern mathematics and to mathematics in the world, written for students of mathematics at the beginning college level.

Koretz, Daniel. *Measuring Up: What Educational Testing Really Tells Us*. Cambridge, MA: Harvard University Press, 2008. A good account, accessible to any intelligent layperson, of the values and pitfalls of educational testing, written by a professor at the Harvard Graduate School of Education who has also taught in public elementary and junior high schools.

Koyré, Alexandre. *From the Closed World to the Infinite Universe.* Baltimore, MD: Johns Hopkins Press, 1957. A philosophically sophisticated account of the emergence of the idea of the infinite universe, addressing religion, philosophy, and astronomy.

Kraut, Richard, ed. *The Cambridge Companion to Plato.* Cambridge: Cambridge University Press, 1992. A valuable and authoritative compendium of recent Plato scholarship.

Laplace, Pierre-Simon. *Philosophical Essay on Probabilities.* New York: Dover, 1995. This is an English translation of Laplace's early 19th-century work on the philosophy of probability, based on lectures he gave at the Ecole Normale in Paris in 1795. The introductory sections are accessible to anybody. Fascinating. This is the cheapest edition; a more scholarly translation is that by Andrew I. Dale, published by Springer in 1995.

Lasserre, François. *The Birth of Mathematics in the Age of Plato.* London: Hutchinson, 1964. A valuable short account of the mathematics of the time of Plato, including a nice chapter on Greek harmonics.

Levinson, Stephen C. "Frames of Reference and Molyneux's Question: Crosslinguistic Evidence." In *Language and Space*, edited by Paul Bloom, Mary A. Peterson, Lynn Nadel, and Merrill F. Garrett, 109–170. Cambridge, MA: MIT Press, 1996. Molyneux's question, posed in the 17th century, was whether a man who has been born blind and who has learned to distinguish and name a globe and a cube by touch would, if somehow enabled to see, be able to distinguish and name these objects simply by sight. Levinson addresses the question more broadly, including whether modes of seeing and representing objects in space are inherent to all humans or are culturally conditioned.

Lloyd, G. E. R. *Adversaries and Authorities: Investigations into Ancient Greek and Chinese Science.* Cambridge: Cambridge University Press, 1996. A scholarly comparison of the different presuppositions of ancient Greek and classical Chinese science and philosophy, focusing, as is clear in the title, on the key way the cultures differed.

Lovejoy, Arthur. *The Great Chain of Being.* Cambridge, MA: Harvard University Press, 1936. The classic work on the principles of plenitude (that the universe is full of the maximal amount of existing things), continuity (that all existing things form a continuum), and unilinear gradation (that all existing things form a hierarchy) and their realization throughout Western thought from Plato's *Timaeus* to 18th-century biology. Especially valuable for its account of the principle of sufficient reason and the infinite universe. A challenging, densely written but supremely important and influential book.

Malthus, Thomas Robert. *An Essay on the Principle of Population, as It Affects the Future Improvement of Society. With Remarks on the Speculations of Mr. Godwin, M. Condorcet, and Other Writers.* London: J. Johnson, 1798. This is the classic work by Malthus, with its pessimistic suggestion that population will eventually outstrip the food supply.

McKeon, Richard. *Introduction to Aristotle.* New York: Modern Library, 1947. Still a good short introduction to Aristotle's major works, though not easy reading.

McKirahan, Richard D., Jr. *Principles and Proofs: Aristotle's Theory of Demonstrative Science.* Princeton, NJ: Princeton University Press, 1992. Excellent scholarly work on the subject for those wishing to pursue the topic further.

Miller, Arthur I. *Einstein, Picasso: Space, Time, and the Beauty That Causes Havoc.* New York: Basic Books, 2002. Parallel biographies of Einstein and Picasso, focusing on the similarities between the way each man confronted the received wisdom of his field and went beyond it, notably with respect to ideas of space and time.

Nadler, Steven. "Baruch Spinoza." In *The Stanford Encyclopedia of Philosophy* (spring 2009 ed.), edited by Edward N. Zalta. http://plato.stanford.edu/archives/spr2009/entries/spinoza/.

Newman, James R., ed. *The World of Mathematics.* New York: Simon and Schuster, 1956. A wonderful 4-volume anthology of writing about mathematics, some technical, much not. Includes all sorts of things, such as W. K. Clifford on non-Euclidean geometry, Lewis Carroll on logic, Nagel and Newman on Gödel's proof, and various short stories on mathematical themes. Hard to find nowadays, but well worth owning if you can get a copy, and well worth seeking out and browsing through in the library.

Newman, James R., and Ernest Nagel. *Gödel's Proof.* 1958. Reprint, New York: New York University Press, 2001. Excellent high-level popularization of Gödel's proof of the incompleteness of mathematics, including enough detail so the assiduous reader can understand the basic idea of the proof.

Niven, Ivan. *Mathematics of Choice: How to Count without Counting.* Washington, DC: Mathematical Association of America, 1965. An excellent introduction to elementary combinatorics. No prerequisite beyond elementary algebra, although it does require close reading. Includes problems to test the reader's understanding. A classic of exposition at the elementary level.

Ortega y Gasset, José. "The Historical Significance of the Theory of Einstein." In *The Modern Theme*, 133–145, 147–152. New York: Harper, 1961. Reprinted in L. Pearce Williams, ed., *Relativity Theory: Origins and Impact on Modern Thought*. Hoboken, NJ: John Wiley, 1968. A thought-provoking essay from the early 1920s, responding to relativity theory as a cultural phenomenon.

Packel, Edward. *The Mathematics of Games and Gambling*. Washington, DC: Mathematical Association of America, 1981. A good introduction to probability theory as applied to games and gambling. Requires algebraic skill and careful reading, but no previous knowledge of probability is needed.

Pascal, Blaise. *Pensées* [*Thoughts*]. 1658 (any edition or translation). This is Pascal's most famous work: a collection of philosophical fragments, some quite polished like "Infinity–Nothing: The Wager." Fascinating reading.

*People v. Collins*, 68 Cal. 2d 319, 438 P.2d 33, 66 Cal. Rptr. 497 (1968). Available at http://isites.harvard.edu/icb/icb.do?keyword=k9840&pageid= icb.page36965&pageContentId=icb.pagecontent90797&state=maximize&vi ew=view.do&viewParam_name=PeoplevCollins.html. This is the court case in which 2 people were convicted on the basis of estimated probabilities that they were the only couple in Los Angeles County with a particular set of personal characteristics. They were, arguably, convicted largely on the basis of a misuse of the multiplication principle.

Peter, Rosza. *Playing with Infinity*. New York: Dover, 1971. A fine introduction to ideas about infinite sets, including the fact that there are infinite sets of different sizes. No mathematical prerequisites are needed, save curiosity, but the book gets into some important and intricate ideas.

Petit, Jean-Pierre. *Here's Looking at Euclid*. Translated by Ian Stewart. Los Altos, CA: William Kaufmann, 1985. A charming introduction to non-Euclidean geometry; includes many cartoons and illustrations as well as clear elementary explanations.

Plato. *Republic*. c. 378 B.C.E. (any edition or translation). Plato's major philosophical work. Addresses the question "What is justice?" and many other questions, ranging from the nature of knowledge and reality to the construction of a stable "ideal" but nondemocratic society. A seminal source for all Western thought and for understanding some of the medieval theology of Christianity, Judaism, and Islam.

*Plato's "Meno."* Translated by George Anastaplo and Laurence Berns. Newburyport, MA: Focus, 2004. Though there are many translations of Plato's *Meno*, this one is notable for its excellent explanations of the mathematics and its systematic series of diagrams of the argument about finding the square with double the area of a given square.

Poincaré, Henri. *Science and Hypothesis.* 1905. Reprint, New York: Dover, 1952. A great mathematician and philosopher of science addresses the relationship among science, nature, and the ways we think. This text is the source of the idea that which axioms we choose for geometry is a matter of convention.

Popkin, Richard H. *The History of Skepticism from Erasmus to Spinoza.* Berkeley: University of California Press, 1979. An excellent work of scholarship, placing people like Montaigne and Descartes in the context of the history of skepticism.

Porter, Theodore. *The Rise of Statistical Thinking, 1820–1900.* Princeton, NJ: Princeton University Press, 1986. An excellent and pioneering history of this topic, especially good for its account of the work of Quetelet. Not easy reading, but worth the effort.

———. "Statistics and Physical Thinking." In *The Modern Physical and Mathematical Sciences*, edited by M. J. Nye, 488–504. Vol. 5 of *The Cambridge History of Science.* Cambridge: Cambridge University Press, 2003. A fine piece of intellectual history, explaining, among other things, how the work of Quetelet interacted with later work like that of Maxwell.

———. *Trust in Numbers: The Pursuit of Objectivity in Science and Public Life.* Princeton, NJ: Princeton University Press, 1995. Sophisticated and challenging, this book argues (with a wealth of examples, especially from the 19th century) that mathematics has been called upon for its ability to bring apparently objective solutions to disputes on important topics where no established authority seems to exist.

Redman, Ben Ray, ed. *The Portable Voltaire.* New York: Viking Press, 1949. A nice collection of Voltaire's writings, including the full text of famous works like *Candide* and many excerpts from Voltaire's *Dictionary.*

Rhodes, Richard. *The Making of the Atomic Bomb.* New York: Touchstone, 1986. A Pulitzer Prize winner, this 886-page book gives a detailed and authoritative account of both the science and technology and the social background of the atomic bomb.

Richards, J. L. "The Geometrical Tradition: Mathematics, Space, and Reason in the Nineteenth Century." In *The Modern Physical and*

*Mathematical Sciences*, edited by M. J. Nye, 449–467. Vol. 5 of *The Cambridge History of Science*. Cambridge: Cambridge University Press, 2003. An excellent account of the topic, especially strong on the response to non-Euclidean geometry in the English-speaking world. First-rate intellectual history.

Robbin, Tony. *Shadows of Reality: The Fourth Dimension in Relativity, Cubism, and Modern Thought*. New Haven, CT: Yale University Press, 2006. For those versed in geometry, this extensively illustrated book is well worth the effort.

Rubin, Ronald. *Silencing the Demon's Advocate: The Strategy of Descartes' Meditations*. Stanford, CA: Stanford University Press, 2008. Describes Descartes' strategy in the *Meditations* as eliminating the grounds for doubt by arguing against a hypothetical adversary. Challenging and sophisticated, but rewarding.

Russell, Bertrand. *A History of Western Philosophy*. 1946. Reprint, London: Routledge, 2004. Highly opinionated yet readable introduction to the important figures in the history of Western philosophy up to the 1940s. Russell is well versed in mathematics, which makes his take on major philosophical figures interesting, but don't take him as the last word, especially on matters involving religion.

Salmon, Wesley. *Logic*. Englewood Cliffs, NJ: Prentice-Hall, 1983. A good—and short—introduction to logic by a distinguished philosopher who knows a lot of science. This is a textbook, so it is systematic and requires careful attention.

Sinaiko, Herman L. "Knowing, Being, and the Community: The Divided Line and the Cave in *Republic*, Books VI and VII." In *Reclaiming the Canon*, 277–299. New Haven, CT: Yale University Press, 1998. A good account of the divided line and the cave, explaining them clearly in Plato's own terms and including an explanation of why the divided line cannot be constructed.

Smith, Gary. *Statistical Reasoning*. New York: McGraw-Hill, 1998. An outstanding and extensive introduction to the use of statistics in politics and economics, with a wealth of examples, including graphs. Written at the level of an introductory college course, presupposing only algebra.

Spinoza, Baruch. *Ethics Demonstrated in Geometrical Order*. 1677 (any edition or translation). This is not light reading, but it is the ultimate expression of Spinoza's rationalism, in the form of a demonstrative science.

Stoppard, Tom. *Arcadia*. London: Faber and Faber, 1993. A modern play (by the author of *Rosencrantz and Guildenstern Are Dead*) that uses the ideas of entropy, chaos theory, and fractals as part of his portrayal of the postmodern human condition. Witty and thought provoking.

Swetz, Frank J., and T. I. Kao. *Was Pythagoras Chinese? An Examination of Right-Triangle Theory in Ancient China*. University Park: Pennsylvania State University Press, 1977. A translation of the section on the "Pythagorean" theorem from the Chinese classic work *Nine Chapters on the Mathematical Art*, with extensive commentary.

Tufte, Edmund. *The Visual Display of Quantitative Information*. Cheshire, CT: Graphics Press, 1983. A beautifully illustrated and well-written book—the best single book on graphs—by a real expert. Highly recommended.

Tversky, Amos, Paul Slovic, and Daniel Kahneman, eds. *Judgment under Uncertainty*. Cambridge: Cambridge University Press, 1982. How people actually reason, as opposed to how one might think they reason or how they ought to reason. A highly original and influential work.

Von Baeyer, Hans Christian. *Maxwell's Demon: Why Warmth Disperses and Time Passes*. New York: Random House, 1998. An attempt to explain thermodynamics and Maxwell's views on thermodynamics to a wider audience. Not an easy read, but quite interesting.

Von Helmholtz, Hermann. "On the Origin and Significance of Geometrical Axioms." 1870. Reprinted in *Popular Scientific Lectures*, 223–249. New York: Dover Publications, 1962. A highly influential essay that argues that we can indeed learn to order our perceptions in non-Euclidean space. Also argues against the view that there is only one a priori intuition of space and in favor of the view that the nature of space is an empirical question.

Wagner, Mark. *The Geometries of Visual Space*. Mahwah, NJ: Lawrence Erlbaum, 2006. By a psychologist well versed in geometry, this book brings together much research about how people perceive space and what the mathematical properties are of the space they perceive.

Wainer, Howard. *Graphic Discovery: A Trout in the Milk and Other Visual Adventures*. Princeton, NJ: Princeton University Press, 2005. A very good book on the history of graphs and on good graphs and deceptive graphs.

Weaver, Warren. *Lady Luck: The Theory of Probability*. Garden City, NY: Doubleday, 1963. An excellent introduction to probability theory with some lively chapters on philosophical issues surrounding probability, this book requires only some knowledge of elementary algebra.

Weizenbaum, Joseph. "Against the Imperialism of Instrumental Reason." In *Computer Power and Human Reason*, 258–280. San Francisco: W. H. Freeman, 1976. Weizenbaum is the computer scientist who designed the Eliza program, which pretended to be a psychiatrist, and who was then horrified that people actually thought a computer could do psychiatry. As a result, he wrote this eloquent book about the limitations of the view that human beings are nothing more than processors of information, arguing instead for ethical and responsible choice.

Wolfson, Harry Austryn. *The Philosophy of Spinoza*. Cambridge, MA: Harvard University Press, 1934. This is the great classic work that analyzes Spinoza's philosophy in the context of medieval, Jewish, and 17th-century sources and gives detailed accounts of all of Spinoza's writings. Tough going but valuable and comprehensive.

Woloshin, Stephen, Lisa M. Schwartz, and H. Gilbert Welch. *Know Your Chances: Understanding Health Statistics*. Berkeley: University of California Press, 2008. A good elementary, popular account, with no mathematical prerequisites at all, of some of the important issues in interpreting medical information presented in statistical form. It has some nice question-and-answer material to help the reader understand.

Wordsworth, William. "The Tables Turned." In *The Collected Poems of William Wordsworth*. Ware, England: Wordsworth Editions, 1998. Often anthologized, this poem presents an antirationalist, Romantic view of nature.

Zamyatin, Yevgeny. *We*. 1952. Reprint, New York: Penguin Group, 1993. An anti-utopian novel in which the instruments of social control are mathematical tables, Zamyatin's book was an important influence on Orwell's *1984*.

# Notes

# Notes

# Notes

# Notes

# Notes